The Chain of Life

Time

A Sketch of the Origin and Succession of Animals and Plants

John William Dawson

Alpha Editions

This edition published in 2024

ISBN : 9789366381718

Design and Setting By
Alpha Editions
www.alphaedis.com
Email - info@alphaedis.com

As per information held with us this book is in Public Domain.
This book is a reproduction of an important historical work. Alpha Editions uses the best technology to reproduce historical work in the same manner it was first published to preserve its original nature. Any marks or number seen are left intentionally to preserve its true form.

Contents

PREFACE..- 1 -
CHAPTER I...- 4 -
CHAPTER II...- 22 -
CHAPTER III..- 42 -
CHAPTER IV...- 83 -
CHAPTER V...- 107 -
CHAPTER VI..- 125 -
CHAPTER VII...- 148 -
CHAPTER VIII..- 165 -
CHAPTER IX..- 184 -
CHAPTER X...- 205 -
CHAPTER XI..- 222 -
FOOTNOTES:..- 235 -

PREFACE.

QUESTIONS as to the origin and history of life are not at the present time answered by mere philosophical speculation and poetical imagining. Such solutions of these questions as science can profess to have obtained are based on vast accumulations of facts respecting the remains of animals and plants preserved in the rocky beds of the earth's crust, which have been successively accumulated in the course of its long geological history. These facts undoubtedly afford the means of attaining to very certain conclusions on many points relating to the history of life on the earth. But, on the other hand, they have furnished the material for hypotheses which, though confidently affirmed to be indisputable, have no real foundation in nature, and are indirectly subversive of some of the most sacred beliefs of mankind.

In these circumstances it is most desirable that those who are not specialists in such matters should be in a position to judge for themselves; and it does not appear impossible in the actual state of knowledge, to present, in terms intelligible to the general reader, such a view of the ascertained sequence of the forms of life as may serve at once to give exalted and elevating views of the great plan of creation, and to prevent the deceptions of pseudo-scientists from doing their evil work. Difficulties, no doubt, attend the attempt. They arise from the number and variety of the facts, from the uncertainties attending many important points, from the new views constantly opening up in the progress of discovery, and from the difficulty of presenting in an intelligible form the preliminary data in biology and geology necessary for the understanding of the questions in hand. In order, as far as possible, to obviate these difficulties, the plan adopted in this work has been to note the first known appearance of each leading type of life, and to follow its progress down to the present time or until it became extinct. This method is at least natural and historical, and has commended itself to the writer as giving a very clear comprehension of the actual state of our knowledge, and as presenting some aspects of the subject which may be novel and suggestive even to those who have studied it most deeply.

In selecting examples and illustrations, the writer has endeavoured to avoid, as far as possible, those already familiar to the general reader. He has carefully sought for the latest facts, while rejecting as unproved many things that are confidently asserted; and has endeavoured to avoid all that is irrelevant to the subject in hand, and to abstain from all technical terms not absolutely essential. In a work at once so wide in its scope, so popular in its character, and so limited in its dimensions, a certain amount of hostile criticism on the part of specialists is to be expected, some portion of it perhaps just, other portions arising from narrow prejudices due to limited lines of study. The writer is willing to receive such comments with attention and gratitude, but he would deprecate the misuse of them in the interest of those coteries which are at present engaged in the effort to torture nature into a confession of belief in the doctrines of a materialistic or agnostic philosophy.

The title of the work was suggested by that of Gaudry's recent attractive book, *Les Enchaînements du Monde animal*. It seemed well fitted to express the connection and succession of forms of life, without implying their derivation from one another, while it reminds us that nature is not a fortuitously tangled skein, and that the links which connect man himself with the lowest and oldest creatures bind him also to the throne of the Eternal.

In the few years that have elapsed since the publication of the first edition of this work, great additions have been made to our knowledge of fossil animals and plants. Many new species have been described, and many new facts have been discovered, respecting species previously known. This rapid progress of discovery has, however, invalidated few of the statements made in the first edition, and has certainly established nothing against the general laws of the succession of life as stated in this work.

Perhaps the most interesting phase of recent discovery is the tracing back of certain forms of life to earlier periods of the earth's geological history. Some of the most recent facts of this kind are the finding, by M. Charles Brongniart, of a fossil insect, allied to the *Blattae* or cockroaches, in the Silurian of Spain, that of true Scorpions in the Upper Silurian of Sweden by Lindström, and in the Upper Silurian of Scotland by Peach, who has also described fossil Millipedes from the Lower Devonian. The tendency of such discoveries is to carry farther back the origin of highly specialised forms of life, and thus to render less probable their origin by any process of gradual derivation.

Other discoveries serve to fill up blanks in our knowledge, and thus to render the geological record less imperfect. Of this kind is the close approximation now worked out in Western America between the end of the reign of the great Mesozoïc reptiles and the beginning of that of the mammals of the Tertiary—a great and abrupt revolution, effected apparently by a *coup de main*. I have myself had opportunity to show that a similarly sharp line separates that quaint old Mesozoïc flora of pines, cycads and ferns, which extends upward into the Lower Cretaceous, from the rich and luxuriant assemblage of broad-leaved trees of modern aspect, which takes its place in the middle part of the same formation.

It is not too much to say that these and similar discoveries, while they serve to bridge over gaps in the succession of organic beings, do not favour the theory of slow modification of types. They rather point to a law of rapid development of new forms under special conditions as yet unknown to science, and this accompanied with the extinction of older species. Recent discoveries also present many remarkable instances of the early introduction of highly specialised types, of higher forms preceding those that are lower in the same class, and of the persistence of certain types throughout geological time without any important change.

J. W. D.

MCGILL COLLEGE.

CHAPTER I.

PRELIMINARY CONSIDERATIONS AS TO THE EXTENT AND SOURCES OF OUR KNOWLEDGE.

It is of the nature of true science to take nothing on trust or on authority. Every fact must be established by accurate observation, experiment, or calculation. Every law and principle must rest on inductive argument. The apostolic motto, "Prove all things, hold fast that which is good," is thoroughly scientific. It is true that the mere reader of popular science must often be content to take that on testimony which he cannot personally verify; but it is desirable that even the most cursory reader should fully comprehend the modes in which facts are ascertained and the reasons on which conclusions are based. Failing this, he loses all the benefit of his reading in so far as training is concerned, and cannot have full assurance of that which he believes. When, therefore, we speak of life-epochs, or of links in a chain of living beings, the question is at once raised—What evidence have we of the succession of such epochs? This question, with some accessory points, must engage our attention in the present chapter.

Geology as a practical science consists of three leading parts. The first and most elementary of these is the study of the different kinds of rocks which enter into the composition of those parts of the earth which are accessible to us, and which we are in the habit of calling the crust of the earth. This is the subject of *Lithology*, which is based on the knowledge of minerals, and has recently become a much more precise department of science than heretofore, owing to the successful employment of the microscope in the investigation of the minute structure and composition of rocks. The second is the study of the arrangement of the materials of the earth on the large scale, as beds, veins, and irregular masses; and inasmuch as the greater part of the rocks known to us in the earth's crust are arranged in beds or strata, this department may be named *Stratigraphy*. A more general name sometimes employed is that of *Petrography*. The third division of geology relates to the remains of animals and plants buried in the rocks of the earth, and which have lived at the time when those rocks were in process of formation. These fossil remains introduce us to the history of life on the earth, and constitute the subject of *Palæontology*.

It is plain that in considering what may be learned as to epochs in the history of life we are chiefly concerned with the last of these divisions. The second may also be important as a means of determining the relative ages of the fossils. With the first we have comparatively little to do.

Previous to observation and inquiry, we might suppose that the kinds of animals and plants which now inhabit the earth are those which have always peopled it; but a very little study of fossils suffices to convince us that vast numbers of creatures once inhabitants of this world have become extinct, and can be known to us only by their remains buried in the earth. When we place this in connection with stratigraphical facts, we further find that these extinct species have succeeded each other at different times, so as to constitute successive dynasties of life. On the one hand, when we know the successive ages of fossil forms, these become to us, like medals or coins to the historian, evidences of periods in the earth's history. On the other hand, we are obliged in the first instance to ascertain the ages of the medals themselves by their position in the successive strata which have been accumulated on the surface. The series of layers which explorers like Schliemann find on the site of an ancient city, and which hold the works of successive peoples who have inhabited the place, thus present on a small scale a faithful picture of the succession of beds and of forms of life on the great earth itself.

Our leading criterion for estimating the relative ages of rocks is the superposition of their beds on each other. The beds of sandstone, shale, limestone, and other rocks which constitute the earth's crust have nearly all been deposited thereon by water, and originally in attitudes approaching to horizontality. Hence the bed that is the lower is the older of any two beds. Hence also, when any cutting or section reveals to us the succession of several beds, we know that fossil remains contained in the lower beds must be of older date.

We can scarcely walk by the side of a stream which has been cutting into its banks, or at the foot of a sea-cliff, or through a road-cutting, without observing illustrations of this. For instance, in the section represented in Fig. 1, we see at the surface the vegetable soil, below this layers of gravel and sand, below this a bed of clay, and below this hard limestone. Of these beds *a* is the newest, *d* the oldest; and if, for example, we should find some marine shells in *d*, some freshwater shells in *c*, bones of land animals and flint arrowheads in *b*, and fragments of modern pottery in *a*, we should be able at once to assign their relative ages to these fossils, and to form some idea of the succession of conditions and of life which had occurred in the locality.

On a somewhat larger scale, we have in Fig. 2 a section of the beds cut through by the great Fall of Niagara. All of these except that marked *a* are very ancient marine rocks, holding fossil shells and corals, but now forming part of the interior of a continent, and cut through by a fresh-water river.

FIG. 1.—Bank of stream or coast, showing stratification.

a, Vegetable soil. *b*, Gravel and sand. *c*, Clays. d, Limestone rock, slightly inclined.

FIG. 2.—Section at Niagara Falls, showing the strata cut through by the action of the Fall. Thickness of beds about 250 feet.

a, Boulder clay and gravel—Post-pliocene.

b, Niagara limestone
c, Niagara shale ⎤ Upper Silurian, with marine shells and corals.
d, Clinton limestone
e, Medina sandstone

In deep mines and borings still more profound sections may be laid open, as in Fig. 3, which represents the sequence of beds ascertained by boring with the diamond drill in search of rock salt near Goderich in Canada. Here we have a succession of 1,500 feet of beds, some of which must have been formed under very peculiar and exceptional conditions. The beds of rock salt and gypsum must have been formed by the drying up of sea-water in limited basins. Those of Dolomite imply precipitation of carbonate of lime and magnesia in the sea-bottom. The marls must have been formed largely by the driftage of sand and clay, while some of the limestone was produced by accumulation of corals and shells. Such deposits must not only have been successive, but must have required a long time for their formation.

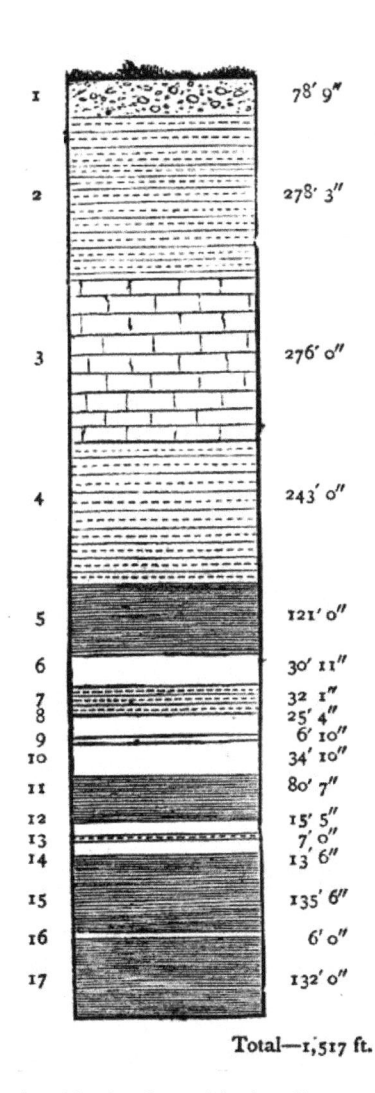

#	Thickness
1	78' 9"
2	278' 3"
3	276' 0"
4	243' 0"
5	121' 0"
6	30' 11"
7	32' 1"
8	25' 4"
9	6' 10"
10	34' 10"
11	80' 7"
12	15' 5"
13	7' 0"
14	13' 6"
15	135' 6"
16	6' 0"
17	132' 0"

Total—1,517 ft.

FIG. 3.—Section obtained by boring with the diamond drill, near Goderich, Ontario, Canada, in the Salina series of the Upper Silurian. From a memoir by Dr. Hunt in the Report of the Geological Survey of Canada for 1876-7.

No. 1, Clay, gravel, and boulders—Post-pliocene.
Nos. 2, 4, 7, 9, 13, Dolomite or magnesian limestone, with layers of marl, limestone, and gypsum.
No. 3, Limestone with corals—*Favosites*, etc.
Nos. 5, 11, 15, 17, Marls with layers of Dolomite and anhydrous gypsum.
Nos. 6, 8, 10, 12, 14, 16, Rock salt.

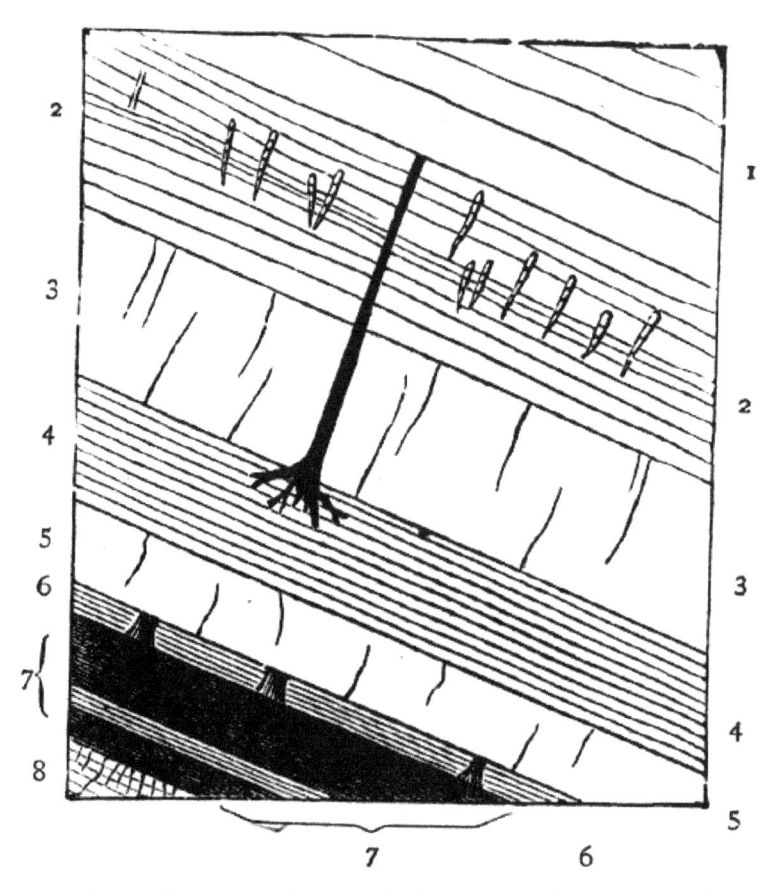

FIG. 4.—Inclined beds, holding fossil plants. Carboniferous. South Joggins, Nova Scotia.

1. Shale and sandstone. Plants with *Spirorbis* attached; rain marks (?).

2. Sandstone and shale, 8 feet. Erect *Calamites*.

⏋ An erect coniferous (?) tree, rooted

3. Gray sandstone, 7 feet.

⎤ on the shale, passes up through 15
⎦ feet of the sandstones and shale.

4. Gray shale, 4 feet.

5. Gray sandstone, 4 feet.

6. Gray shale, 6 inches. Prostrate and erect trees, with rootlets, leaves, *Naiadites*, and *Spirorbis* on the plants.

7. Main coal-seam, 5 feet coal in two beds.

8. Underclay, with rootlets.

In [Fig. 4](#) we have a bed of coal and its accompaniments. The coal itself was produced by the slow accumulation of vegetable matter on a water-soaked soil, and this was buried under successive beds of sand and clay, now hardened into sandstone and shale, some of the beds holding trees and reed-like plants, which still stand on the soils on which they grew, and which must have been buried in sediment deposited in inundations or after subsidence of the land. In this section we may also observe that the beds are somewhat inclined; and that this is not their original position is shown by the posture of the stems of trees, once erect, but now inclined with the beds. This leads to a consideration very important with reference to our present subject; namely, that as our continents are mostly made up of beds deposited under water and afterwards elevated, these beds have in this process experienced such disturbances that they rarely retain their horizontal position, but are tilted at various angles. When we follow such inclined strata over large areas, we find that they undulate in great waves or folds, forming what are called anticlinal and synclinal lines, and that the irregularities of the surface of the land depend to a great extent on these undulations, along with the projection of hard beds whose edges protrude at the surface. In point of fact, as shown in [Fig. 5](#), mountain ranges depend on these crumplings of the earth's crust; and the primary cause of these is probably the shrinkage of the mass of the earth owing to contraction in cooling. When the disturbances of beds are extreme, they often cause intricacies of structure difficult to unravel; but when of moderate extent they very much aid us in penetrating below the surface, for we can often see a great thickness of beds rising one from beneath another, and can thus know by mere superficial examination the structure of the earth to a great depth. It thus happens that geologists reckon the thickness of the stratified deposits of the crust of the earth at more than 70,000 feet, though they cannot penetrate it perpendicularly to more than a fraction of that depth. The two sections, Figs. 6 and 7, showing the sequence of beds in England and in the northern part of North America, will serve, if studied by the reader, to show how, by merely travelling over the surface and measuring the upturned edges of beds, many thousands of feet of deposits may be observed, and their relative ages distinctly ascertained.

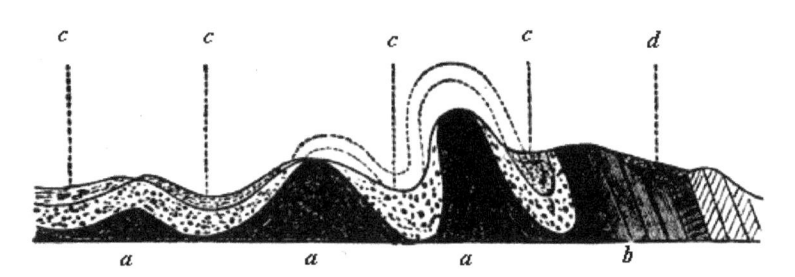

FIG. 5.—Ideal section of the Apalachian Mountains showing folding of the earth's crust.

a, Anticlinal axes. *b*, Overturned strata. *c*, Synclinals. *d*, Unconformable beds.

In studying any extensive section of rock we find that its members may more or less readily be separated into distinct groups. Sometimes these are distinguished by what is termed unconformability, that is, the lower series has been disturbed or inclined before the upper has been deposited upon it. This is seen on a grand scale in the section Fig. 7, in the case of the Laurentian and Cambrian formations, and on a smaller scale in Fig. 8 in the unconformable superposition of Devonian conglomerate on Silurian slates at St. Abb's Head. In the last section it is quite evident that the beds of the lower series have been bent into abrupt folds and worn away to a considerable extent before the deposition of the overlying series. In such a case we know not merely that the upper series is newer than the lower, but that some considerable time must have elapsed after the deposition of the one before the other was laid down; and we are not surprised to find that the fossils in the groups thus unconformable to each other are very different.

But even when the beds are conformable, they can usually be separated into groups, depending upon differences of mineral character, or changes which have occurred in the mode of deposition. One group of beds, for example, may be largely composed of limestone, another of sandstone or shale. One group may be distinguished by containing some special mineral, as, for example, rock salt or coal, while others may be destitute of such special minerals. One group may show by its fossils that it was deposited in the sea, others may be estuarine or lacustrine. Thus we obtain the means of dividing the rocks of the earth into groups of different ages, known as "Formations," and marking particular periods of geological time. By tracing these formations from one district or region to another, we learn the further truth that the succession is not merely local, but that, though liable to variation in detail, its larger subdivisions hold so extensively that they may be regarded as world-wide in their distribution.

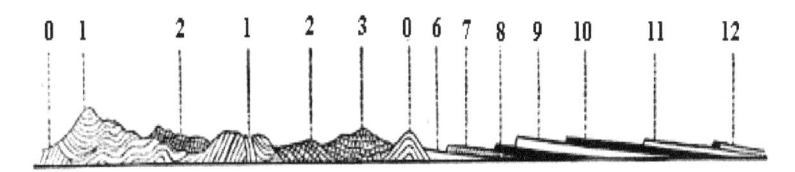

FIG. 6. Generalised section across England from Menai Straits to the Valley of the Thames.—After Ramsay.

0 Huronian? or Laurentian? 1 Cambrian and Lower Silurian. 2 Upper Silurian. 3 Devonian. 6, 7, 8 Trias and lias. 9 and 10 Jurassic. 11 Cretaceous. 12 Eocene.

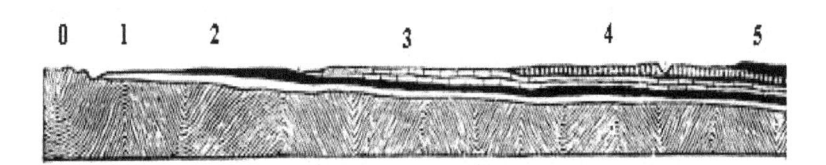

FIG. 7.—Generalised section from the Laurentian of Canada to the coal-field of Michigan.

0 Laurentian (the Huronian is absent in the line of this section). 1 Cambrian. 2 Lower Silurian. 3 Upper Silurian. 4 Devonian. 5 Carboniferous.

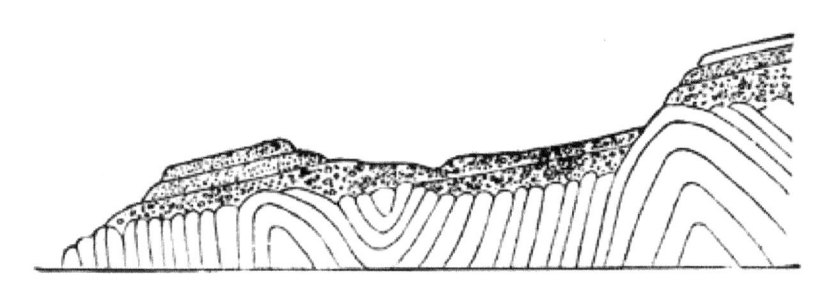

FIG. 8.—Unconformable superposition of Devonian conglomerate on Silurian slates, at St. Abb's Head, Berwickshire.—After Lyell.

Putting together the facts thus obtained, we can frame a tabular arrangement of the earth's strata, as in the table prefixed to this chapter; and when we add the further discovery, very early made by geologists, that the successive formations differ from each other in their fossil remains, we have the means of recognising any particular formation by its fossils, even when the stratigraphical evidence may be obscure or wanting. Thus our knowledge of Epochs of Life, and indeed of the whole geological history of the earth, is based on the superposition of beds in the earth's crust, and on the diversity of fossil remains in the successive beds so superimposed on each other; and it is on these grounds that we are enabled to construct a Table of Geological Formations representing the whole series of beds as far as known, with the characteristic groups of fossils of each period. Here I might close these preliminary considerations, but there are a few accessory questions, important to our clear comprehension of the subject, which may profitably occupy our attention for a short time.

One of these relates to the absolute duration of the time represented by the geological history of the earth. Such estimates as our present knowledge enables us to form are very indefinite. Whether we seek for astronomical or geological data, we find great uncertainty. To such an extent is this the case, that current estimates of the time necessary to bring the earth from a state of primitive incandescence to its present condition have varied from fifteen millions of years to five hundred millions. Of the various modes proposed, perhaps the most satisfactory as well as instructive is that based on the rate of denudation of our present continents, as indicated by the amount of sediment carried down by great rivers. The Mississippi, draining a vast and varied area in temperate latitudes, is washing away the American land at the rate of one foot in 6,000 years. The Ganges, in a tropical climate and draining many mountain valleys, works at the rate of one foot in 2,358 years. The mean of these two great rivers would give one foot in 4,179 years, at which rate our continents would be levelled with the waters in about six millions of years. But the land has been in process of renewal as well as of waste in geological time; and a better measure will be afforded by the amount of beds actually deposited. The entire thickness of all the stratified rocks of Great Britain has been calculated by Ramsay at 72,000 feet. Now, if we suppose the waste in all geological time to have been on the average the same as at present, and that this material has been deposited to the thickness of 72,000 feet on a belt of sea margin 100 miles in width, we shall have about 86 millions of years as the time required.[1] This has the merit of approximating to Sir William Thomson's calculation, based on the rate of cooling of the earth, that a minimum of 100 millions of years may represent the time since a solid crust first began to form. As it is more likely that the rate of denudation has on the average been greater in former geological periods than at present, we may perhaps estimate fifty or sixty millions of years as the time required for the accumulation of all our formations. Some geologists object to this as too little, but in this some of them are influenced by the exigencies of theories of evolution, and others appear to have no adequate conception of the vast lapse of time represented by such numbers, in its relation to the actual rates of denudation and deposition.

It should be mentioned here, however, that, on certain theories now somewhat generally accepted, respecting the nature and source of solar heat, the absolute duration of geological time would be much reduced below the estimate of Sir Wm. Thomson. Prof. Tait has based on such data an estimate of fifteen millions of years. Prof. Simon Newcomb says that "on the only hypothesis science will now allow us to make respecting the source of the solar heat" (the gravitation hypothesis of Helmholtz) "the earth was, twenty millions of years ago, enveloped in the fiery atmosphere of the sun." Dr. Kirkwood has called attention to these results in connection with the planetary hypothesis of La Place, in the *Proceedings of the American Philosophical Society*.[2] Should such views prove to be well-founded, geological calculations as to the time required for the successive formations may have to be revised.

If now we attempt to divide this time among the formations known to us, according to their relative thicknesses, we have, according to an elaborate estimate of Professor Dana, the time ratios of 12, 3, and 1 for the Palæozoic, Mesozoic, and Cainozoic periods respectively. Taking the whole time since the beginning of the Cambrian as forty-eight millions of years, we should thus have for the Palæozoic thirty-six millions, for the Mesozoic nine, and for the Tertiary three. Another calculation, recently made by Professors Hull and Haughton, gives the following ratios:—

Azoic   34·3 per cent.

Palæozoic   42·5   ”

Mesozoic and Cainozoic   23·2   ”

This calculation is, however, based on the absolute thickness of the several series as ascertained in Great Britain, without reference to the nature of the beds, as indicating different rates of accumulation. Under either estimate it will be seen that the Palæozoic time greatly exceeds the Mesozoic and Cainozoic together, and consequently that changes of life seem to have proceeded at an accelerated rate as time wore on.

Another inquiry of some importance relates to the manner of preservation of fossils, and the extent to which they constitute the material of rocks. This inquiry is doubly important, as it bears on the genuineness of fossil remains, and on the means we have of understanding their nature.

Some rocks are entirely made up of matter that once was alive, or formed part of living organisms. This is the case with some limestones, which consist of microscopic shells, or of larger shells, corals, and similar calcareous organisms, either entire or broken into fragments and cemented together with pasty or crystalline limestone filling their interstices. This may be seen in Fig. 9, which represents a magnified slice of a Silurian limestone. Coal in like manner consists of carbonised vegetable matter, retaining more or less perfectly its organic structure, and sometimes even the external forms of its constituent parts. More frequently, fossils are dispersed more or less sparsely through the substance of beds composed of earthy matter; and they have usually been more or less affected by chemical changes, or by mechanical pressure, or are mineralised by different substances which have either filled their pores by infiltration or have more or less completely replaced their substance. Of course, as a rule, the softer and more putrescible organic matters have perished by decay, and it is only the harder and more resisting parts that remain. Even these have often yielded to the enormous pressure to which they have been subjected, and if at all porous, have been changed by the slow action of percolating water charged with various kinds of mineral matter in solution.

FIG. 9.—Section of Trenton limestone, magnified, showing that it is composed of fragments of corals, crinoids, and shells. Montreal.

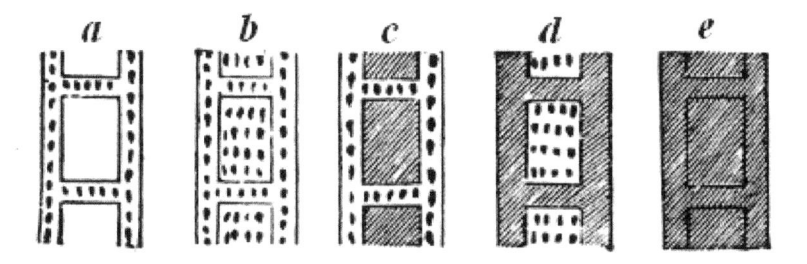

FIG. 10.—Diagram showing different state of fossilisation of a cell of a tabulate coral (Dawson's *Dawn of Life*).

a Natural condition, wall calcite cell empty. *b* Wall calcite, cells filled with the same. *c* Walls calcite, cells filled with silica or a silicate. *d* Wall silicified, cells filled with calcite. *e* Wall silicified, cell filled with silica.

It thus happens that many fossils are infiltrated with mineral matter. Wood, for example, may have the cavities of its cells and vessels filled with silica or silicates, with sulphide or carbonate of iron, or with limestone, while the woody walls of the cells may remain either as coaly matter or charcoal. I have often seen the microscopic cells of fossil wood not only filled in this way, but presenting under a high power successive coats of deposit, like the banded structure of an agate.

In some cases not only are the pores filled with mineral matter, but the solid parts themselves have been replaced, and the whole mass has actually become stone, while still retaining its original structure. Thus silicified wood is often as hard and solid as agate, and under the microscope we see that the wood has entirely perished, and is represented by silica or flint, differing merely in colour from that which fills the cavities. In this case we may imagine the wood to have been acted on by water holding in solution silica, combined with soda or potash, in the manner of what is termed soluble glass. The wood, in decay, would be converted into carbon dioxide, and this as formed would seize on the potash or soda, leaving the silica in an insoluble state, to be deposited instead of the carbon. Thus each particle of the carbon of the wood, as removed by decay, would be replaced by a particle of silica, till the whole became stone. By similar chemical changes corals and shells are often represented by silica, or by pyrite, which has taken the place of the original calcareous matter; and still more remarkable changes sometimes occur, as when the siliceous spicules of sponges have been replaced by carbonate of lime. The organic matter present in the fossils greatly promotes these changes, by the substances produced in its decay, and thus it often happens that the shells, corals, etc., contained in limestone have been replaced by flint, while the inclosing limestone is unchanged. Fig. 10 shows the various conditions which a coral may assume under these different modes of treatment.

The substance of a fossil may be entirely removed by decay

FIG. 11.—Cast of erect tree (*Sigillaria*) in sandstone, standing on a small bed of coal, South Joggins, Nova Scotia (Dawson's *Acadian Geology*). or solution, leaving a mere mould representing its external form, and this may subsequently be filled with mineral matter, so as to produce a natural cast of the object. This is very common in the case of fossil plants; and large trunks of trees may sometimes be found represented, as seen in Fig. 11, by stony pillars retaining nothing of the original wood except perhaps a portion of the bark in the state of coal. It sometimes happens that the substance of fossils has been removed, leaving mere empty cavities, sometimes containing stony cores representing the internal chambers of the fossils. Again, calcareous fossils imbedded in hard rocks are often removed by weathering, leaving very perfect impressions of their forms. For this reason the fossil remains contained in some hard resisting rocks can be best seen as impressed moulds on the weathered surfaces.

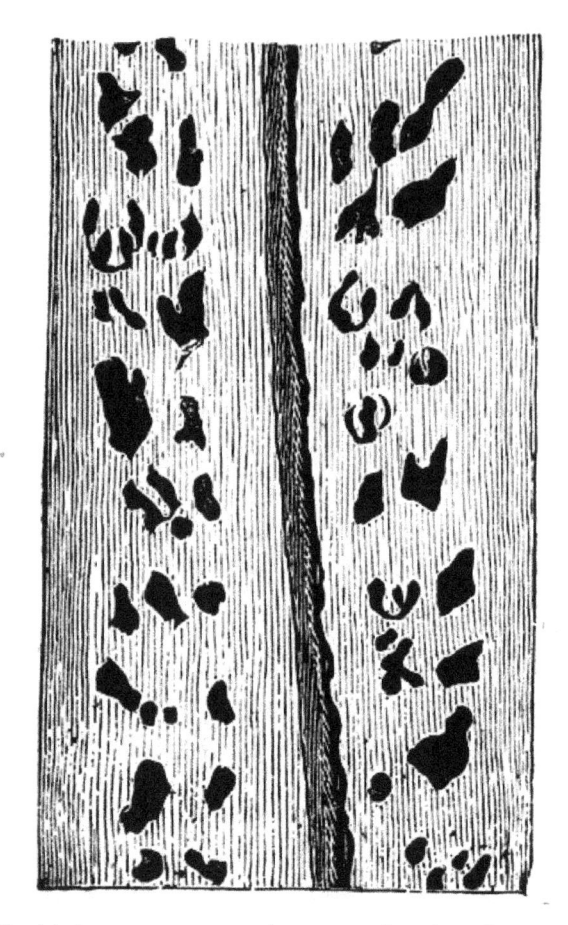

FIG. 12.—*Protichnites septem-notatus.* A supposed series of crustacean footprints made in sand, now hardened into sandstone. Cambrian.—After Logan.

Lastly, we sometimes have impressions or footprints representing the locomotion of fossil animals, rather than the fossils themselves. In this way some extinct creatures are known to us only by their footsteps on sand or clay, once soft, but now hardened into stone; and in the case of some of the lower animals the trails thus made are often not easily interpreted (Figs. 12, 12a). It has been found that even sea-weeds drifted by the tide make impressions of this kind, which, when they occur in old rocks, are very mysterious. Even rain-drops are capable of being permanently impressed on rocks, and constitute a kind of fossils. Besides these we have many kinds of imitative markings which simulate fossils, as those of concretions or nodules, which are often very fantastic in shape, those of dendritic crystallisation giving moss-like forms, and the complicated tracery produced on muddy shores by the little rills of water which follow the receding tide (Fig. 13). Such things are often mistaken by the ignorant for fossil remains, but are easily distinguished by a practised eye.

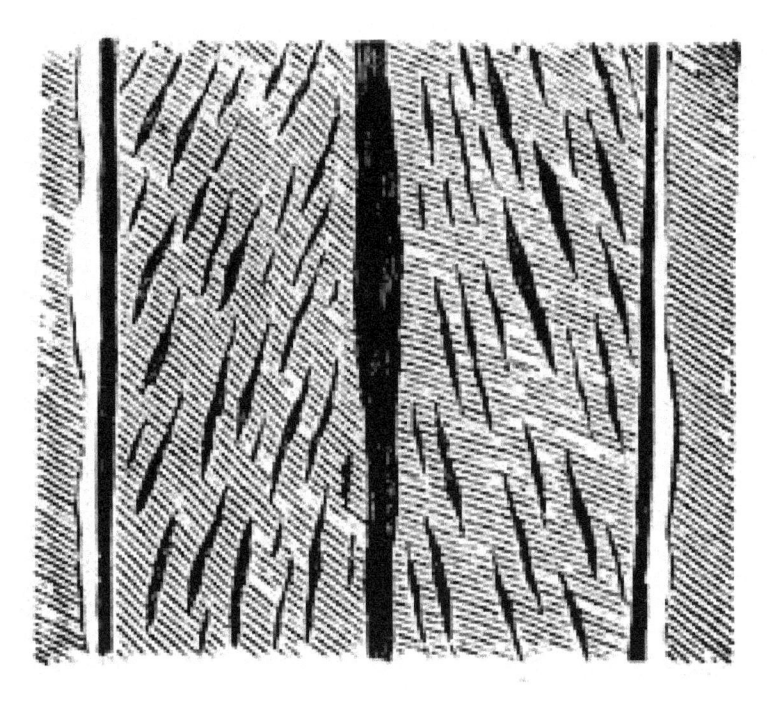

FIG. 12a.—Footprints of modern *Limulus*, or king-crab, in the sand, which enable us to interpret those in Fig. 12.

The reader who has followed these, perhaps somewhat dry, details, will be rewarded for his patience by having some conception of the conditions in which we find fossil remains, and of the evidence by which we can refer these to different periods in the history of the earth.

FIG. 13.—Current markings on shale, resembling a fossil plant. Reduced from a photograph (Dawson's *Acadian Geology*).

Carrying this knowledge with us, and at the same time glancing at the table of successive formations prefixed to this chapter, we shall be prepared, without any additional geological study, to understand the statements to be made in the following chapters, and to appreciate the actual nature of the succession of life in so far as it is at present known.

MAGNIFIED AND RESTORED SECTION OF A PORTION OF EOZOON CANADENSE.

The shaded portions show the animal matter of the Chambers, Tubuli, Canals, and Pseudopodia; the unshaded portions the calcareous skeleton.

CHAPTER II.

THE BEGINNING OF LIFE ON THE EARTH.

THE day must have been when the first living being appeared for the first time on our planet. Was it plant or animal? or a generalised organism uniting in some mysterious way the properties and powers of two kingdoms of nature, now so distinct, and even contrary to each other in their manifestations? Did it appear suddenly, or was it slowly evolved from dead matter by some process in which the albuminous or protoplasmic matter, which we know forms the basal substance of living beings, was first produced and then endowed with life? Did the first living being appear in a mature state, or was it merely a germ from which the mature individual could be produced? These are questions which science in its present state has no means of answering. We do not know any process by which the ingredients of protoplasm can be combined so as to produce that substance without a previous living being. We do not know what molecular differences may exist between dead albumen and that which we see growing and moving and instinct with life; still less do we know how to set up or establish these differences. We do not know the precise nature or relation to other forces of the energy which actuates living organisms. In our experience the simplest creatures that have life spring from previous germs, themselves the products of previous generations of living beings. Thus we are in the presence of great mysteries which it might be impossible for us to solve, even if we were permitted to visit some new planet on which the dawn of life was breaking.

Some things, however, we can infer as to the conditions of the introduction of life.

First, there is every reason to believe that the earth we inhabit was once a glowing, incandescent mass, condensing from a vaporous condition, and quite unfit for the abode of living beings, and which, even if in some previous state its materials had constituted the mass of an inhabited world, must have lost every trace of any living germ in the fervent heat to which it had been subjected. There must, therefore, have been in some way an absolute creation or origination of life and organisation.

Secondly, we may infer that in the earlier stages of the earth, when it was perhaps wholly or almost entirely covered with the waters, when it was still uniformly warmed with its own internal heat, when it was surrounded with a pall of dense vapours preventing radiation, and nursing its heat within itself, though in a condition entirely unsuited to the higher forms of life, it may have presented circumstances more favourable to the origination and multiplication of living beings of low organisation than at any subsequent time. This incubation of creative power in the vaporous mantle over the primæval ocean was a favourite imagination of old thinkers, and is not obscurely hinted at in the Book of Genesis. It has been revived and much insisted on by evolutionists in our own time, though it has no certain foundation in scientific observation or experiment.

Thirdly, from the fact that plant-life alone has the power of subsisting on inorganic matter, and that plants furnish all the nourishment of animals, we may fairly infer that the life of the plant preceded that of the animal. It has, indeed, been suggested that some of the humbler forms of life may combine in a rude and simple way enough of the powers of the plant and the animal to enable them to bridge over the double gap between the animal and the plant, and the animal and the mineral, or that such creatures may in their early stages carry on vegetable functions, and in their later those of the animal. It is theoretically possible that life may have begun with such creatures, which some of the results of microscopical research would lead us to believe still exist. It is, however, on the whole more probable that simple plants first existed, and furnished pabulum to animals of low grade introduced almost contemporaneously.

Fourthly, all our knowledge of the succession of life leads us to believe that it was not the higher plants and animals that first sprang into existence from the teeming earth, but creatures of low and humble organisation, suited to the then immature and unfinished condition of the planet. It is also in accordance with the amazing fecundity of the seas in all geological periods in these lower forms of life, to suppose that the earliest living things originated in the waters, and that the plants and animals of the land are of later date.

Do we know anything from actual observation of this earliest population of the world? Such knowledge we can hope to acquire only by studying the oldest formations known to us; and these, it must be confessed, exist in a state so highly crystalline, and so much affected by internal heat, by mechanical pressure, and by movement, as to render it little likely that organic remains should be preserved in them in a state fit for recognition.

In many parts of the world, and notably in Canada and Scandinavia, as well as in Wales, Scotland, and Bavaria, the older Palæozoic rocks, the lowest containing plants in great abundance, rest on still older crystalline beds, which have become hard and crystalline in pre-Palæozoic times, and have contributed sand and pebbles to the succeeding very ancient deposits. These old rocks—the Eozoic series of our table—may be grouped in two great systems, the Laurentian and Huronian (Fig. 14). The former may be conveniently divided into three members: First, the Bojian, or Ottawa gneiss, consisting of stratified granite rocks, usually of a red colour, and of very great thickness. This contains, so far as known, no limestone, and has afforded as yet no trace of fossils. Secondly, the Middle Laurentian, the greater part of which consists of gneiss, but containing important beds of other rocks, as quartzite, iron ore, and limestone. It is in this series that we have the first evidence of life, and it is here also that we find the greatest abundance of carbon, in the form of graphite or plumbago, and also large quantities of calcium phosphate, or bone earth. Thirdly, the Upper Laurentian or Norian series. This consists in great part of Labadorite, or lime feldspar, but has also beds of ordinary gneiss, limestone, and iron ore.

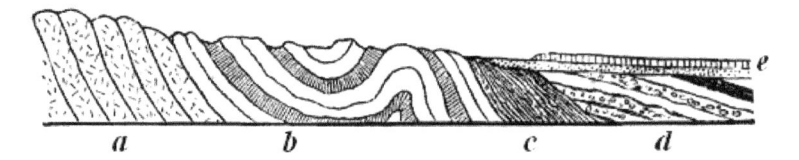

FIG. 14.—Ideal section, showing the relations of the Laurentian and Huronian.

a, Lower Laurentian. *b*, Middle Laurentian. *c*, Upper Laurentian. *d*, Huronian. *e*, Cambrian and Silurian.

The latter, the Huronian, is much less crystalline, and is divisible into two series—the Lower Huronian, which includes many beds of volcanic origin, and the Upper Huronian, which has afforded some obscure fossils. The Huronian was first recognised by Sir W. E. Logan in Canada, but corresponding rocks exist in Europe. The Pebidian series of Hicks in Wales is probably of this age.

It is likely that much of the present appearance and condition of the most ancient rocks may be attributed to metamorphism, that is, to the slow baking under the influence of heat, heated water, and pressure, to which they have been subjected in the lower parts of the earth's crust, when buried deeply under newer deposits. It is also true, however, as Dr. Sterry Hunt has pointed out in detail, that they present mineral characters which show a mode of deposition different from that which has prevailed subsequently, and probably indicating great ejections of heated mineral matter into the primitive ocean, and comparatively little of that deposit therein of mere sand and clay which has prevailed in subsequent geological periods. In short, these rocks have an unmistakably primitive aspect, distinguishing them from those of later times, and conveying the impression that they approach at least to the records of that time when a heated ocean first rested on the thin and recently solidified crust of our planet. If this is really the case, then our Lower Laurentian—hard, compact, destitute of limestone, and composed of material which may be little else than the *débris* of products of internal heat merely spread out into bedded forms by water—may represent a time when no living thing as yet tenanted the waters; and the dawn of life may have appeared in that period when the Middle Laurentian beds were laid down. Here at least we find two kinds of evidence pointing to the existence of certain forms of life in the waters.

The first depends on the mineral character of the beds themselves. This formation holds several very thick beds of limestone. Now although this kind of rock may, under certain circumstances, be deposited directly from solution in water, it is not ordinarily so deposited, but more usually through the agency of living beings inhabiting the waters, and forming their skeletons or hard parts of limestone derived from the water, usually through the medium of humble forms of plant life. In this way are formed reefs of coral and beds of shells and of chalky ooze, all composed of material once constituting the skeletons of animals. The study of limestones of all geological ages shows that this has been the usual mode of their formation. If the Laurentian limestones had a similar origin, the seas of that period must have swarmed with animals having calcareous coverings; and the study of more modern limestones which have become highly crystalline shows that it is quite possible that the forms and structures of these organisms may have been obliterated.

Again, the Middle Laurentian abounds in carbon or coaly matter. True, this is in the form of graphite or plumbago, but this condition may be a result of metamorphism; and we know that the carbon of coal-beds and bituminous shales of much more modern times has been altered into graphite. Further, the graphite occurs in the way in which we should expect it to occur if of organic origin. It is found disseminated in the limestone, just as bituminous matter is found in unaltered rocks of this kind. It is found interlaminated with gneiss, as carbonaceous and bituminous matters are found in the shales of the ordinary fossiliferous rocks, where these substances are known to be of organic origin. The graphite also occurs in a very pure form in irregular veins, just as in some bituminous formations the rock oil, oozing into fissures, has been hardened into asphalt or coaly matter.[3]

To these facts may be added the presence of thick beds and veins of iron ore and of apatite or calcium phosphate (bone earth). Both of these substances occur in a disseminated state in nearly all rocks, but they are concentrated into definite deposits by the action of life. Iron is usually dissolved out and redeposited by acids produced in the decay of vegetable matter, as we see in the clay ironstones of the coal formation and in bog-iron ores. Calcic phosphate is taken up by many animals, and forms their shells or skeletons, and on their death is deposited in beds on the sea-bottom, sometimes to a very considerable extent.

The concurrence of all these phenomena in the Middle Laurentian may be held to afford a strong presumption that, could we discover these rocks in an unaltered state, we should find the limestones filled with marine fossils and the graphite showing the forms or structure of plants. The only startling feature in this conclusion is, that if we admit it, we must also admit that life was developed in the Laurentian time in an exuberance not surpassed, if equalled, in any subsequent period. Still, there is nothing incredible in this, for if the forms of life were few and low, their increase may have been rapid, because unchecked; and they no doubt found in the ancient seas a surplusage of material on which to feed and with which to construct their skeletons. Dr. Hunt has estimated that the amount of carbon now sealed up as coaly matter would, if diffused in the atmosphere as carbon dioxide, afford 600 times the quantity of that gas at present floating in the air. A still more vast amount is sealed up in the limestone of the several geological formations. The same chemist has shown that the quantity of lime held in solution in the ocean must have been much greater in Laurentian times than at present. These facts at least allow us to suppose that in the Eozoic times there were great supplies of carbon and of lime available to such creatures of low organisation as were capable of profiting by them; and we have no reason to doubt that there may have been plants and animals so constituted as to flourish in conditions of this kind, in which perhaps scarcely any modern species could exist.

These probabilities have caused geologists anxiously to search for any traces of fossil organic remains in the old Laurentian rocks; and they have been rewarded by the discovery of one species, *Eozoon Canadense*, still often referred to as only a problematical fossil; but this arises to a large extent from the prevalent want of knowledge sufficient to appreciate the evidence for its organic character. This being once admitted, we have in the existence of *Eozoon* alone a sufficient cause for the accumulation of much of the Laurentian limestone, though there is reason to believe that it was not the only inhabitant of those ancient seas.

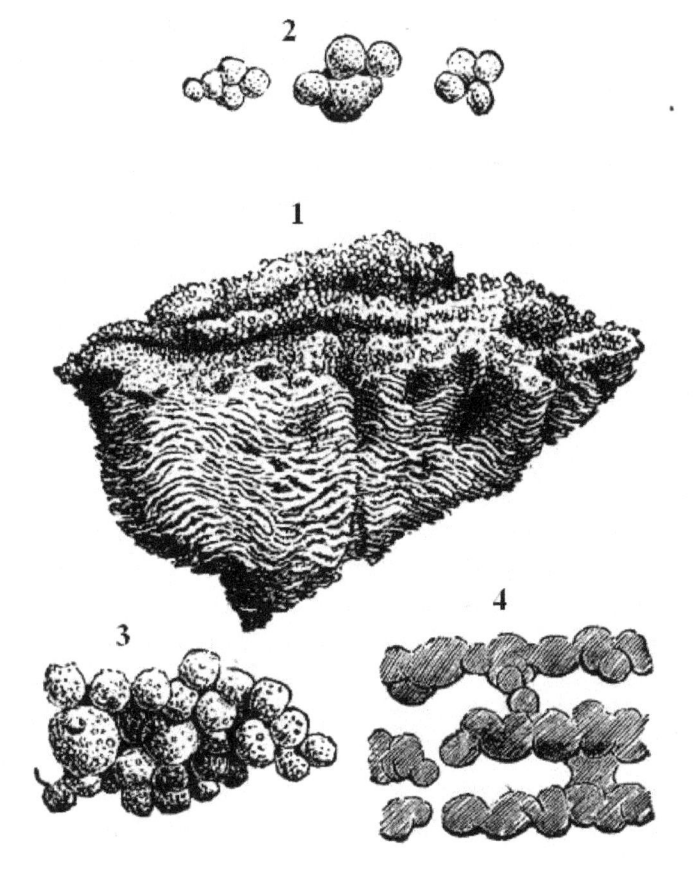

FIG. 15 (Nos. 1 to 4).—Small weathered specimen of *Eozoon*. From Petite Nation.

1, Natural size; showing general form, and acervuline portion above and laminated portion below. 2, Enlarged casts of cells from upper part. 3, Enlarged casts of cells from the lower part of the acervuline portion. 4, Enlarged casts of sarcode layers from the laminated part.

The best specimens of *Eozoon* occur as rounded, flattened, or more or less irregular lumps or masses in certain layers of the Laurentian limestone. When weathered on the surface of the rock, these lumps show a regular concentric lamination, caused by thin fibres of limestone, alternating with other mineral substances, filling up the spaces between them. When these intervening layers are composed of such minerals as Serpentine, Loganite, Pyroxene, or Dolomite, which are more resisting than the limestone, they project when weathered, or when the limestone is etched by an acid, so as to show the lamination very distinctly. At the lower surface of the masses the layers are seen to be thicker than they are above, and in perfect specimens they are seen toward the surface to break up into small rounded vesicles of calcite, like little bubbles, which constitute the so-called acervuline condition of *Eozoon* (Fig. 15, No. 2). Slices of the fossil etched with an acid show these appearances very perfectly, and can even be printed from, so as to present perfect nature-prints of the structure (Fig. 16).

FIG. 16.—Nature-printed specimen of *Eozoon* slightly etched with acid. It shows the lamination, and at one side fragmental *Eozoon* (*Life's Dawn on Earth*).

On etching a small fragment or slice with very dilute acid, so as to dissolve away the calcite slowly, if the specimen be well preserved, we find that the calcite layers have a very curious structure. This is indicated by the appearance of little white or transparent threads of Serpentine, Dolomite, or Pyroxene, which ramify throughout the substance of the limestone layers, and are left intact when they have been dissolved. These little processes must originally have been pores in the limestone layers, which have been filled with the substance which constitutes the alternate laminæ. In addition to this, if we use a somewhat high microscopic power, and especially if we study the structures as seen in thin transparent slices, we can perceive a still finer tubulation along the sides of the calcite layers, represented by extremely minute parallel rods of mineral matter (Figs. 17, 18).

Now if we regard these structures as those of an infiltrated fossil, as described in last chapter, their interpretation will not be difficult. The original organism was composed of calcareous matter in thin concentric laminæ, connected with each other by pillars and plates of similar material. Between these laminæ was lodged the soft, jelly-like substance of a marine animal, growing by the addition of successive layers, each protected by a thin calcareous crust. The layers were originally traversed by very numerous parallel tubuli, permitting the soft protoplasm to penetrate them; and when, in the progress of growth, it was necessary to strengthen these layers, they were thickened by a supplemental deposit traversed by larger and ramifying canals. When the animal was dead, and its soft parts removed by decay, the chambers between the laminæ, as well as the minute canals and tubuli, became infiltrated with mineral matter, in the manner described in the last chapter, and when so preserved became absolutely imperishable under any circumstances short of absolute fusion.

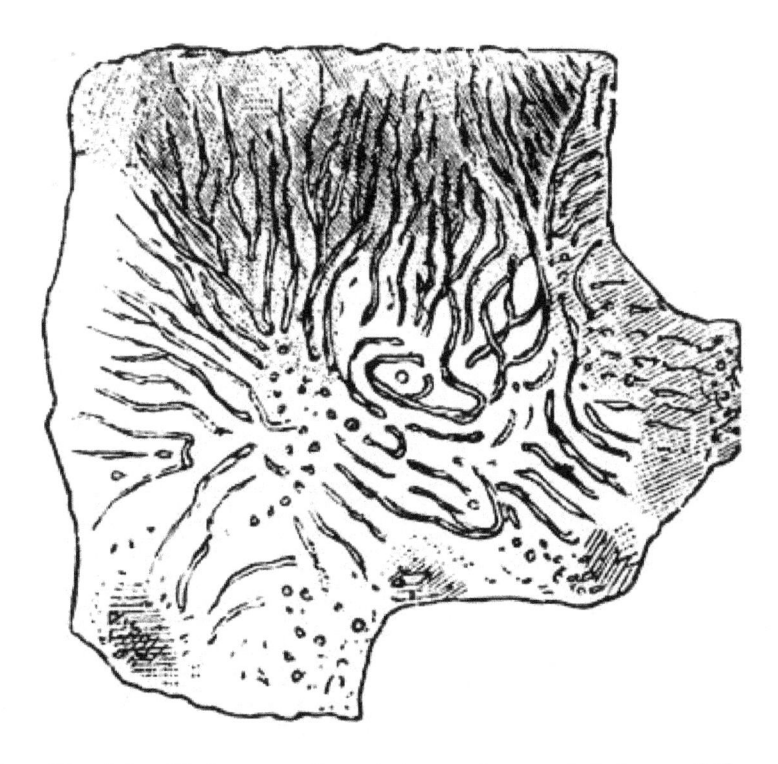

FIG. 17.—Magnified group of canals in supplemental skeleton of *Eozoon*.

Taken from the specimen in which they were first recognised (*Life's Dawn on Earth*).

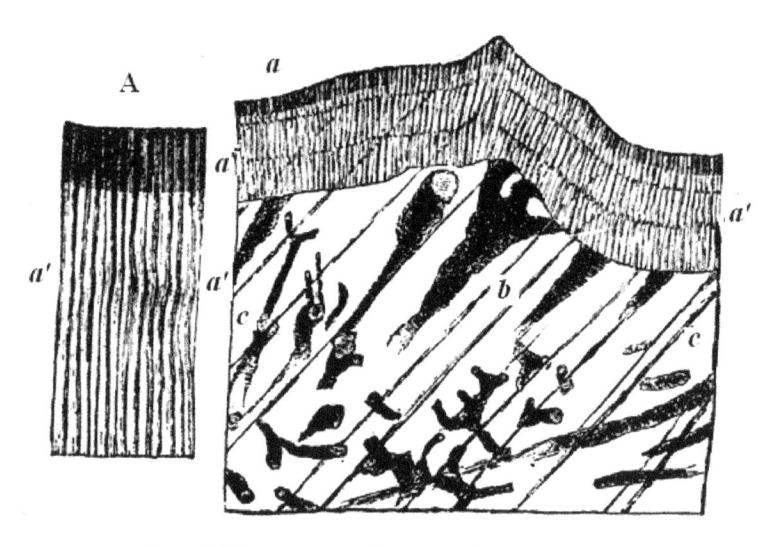

FIG. 18.—Portion of *Eozoon* magnified 100 diameters, showing the original cell-wall with tubulation, and the supplemental skeleton with canals.—After Carpenter.

a, Original tubulated wall or "Nummuline layer." More magnified in Fig. A.
b, *c*, Intermediate skeleton, with canals.

This interpretation leads to the conclusion, at which I arrived from the study of the first well-preserved specimen ever submitted to microscopic examination, that the animal which produced the calcareous skeleton of *Eozoon* was a member of that lowest grade of Protozoa known as Foraminifera; and which, after living through the whole of geological time, still abound in the sea. The main differences are, that *Eozoon* presents a somewhat generalised structure, intermediate between two modern types, and that it attained to a gigantic size compared with most of these organisms in later periods. How near it approaches in structure to some modern forms may be seen by comparison of the recent species represented in Fig. 19, in which the parts corresponding to the chambers, laminæ, tubuli, and canals of *Eozoon* can be readily distinguished.

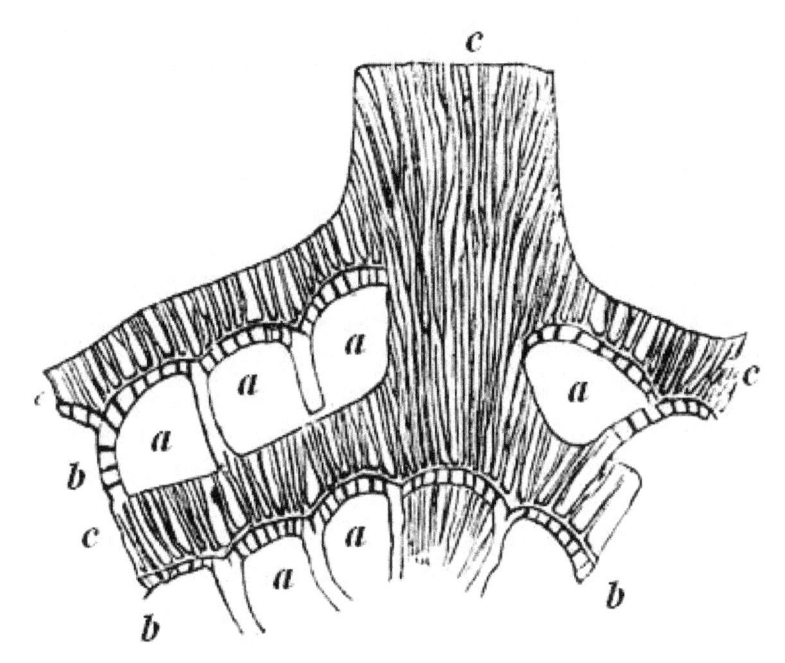

FIG. 19.—Magnified portion of shell of *Calcarina*.—After Carpenter.

a, Cells. *b*, Original cell-wall with tubuli. *c*, Supplementary skeleton with canals.

The modern animals of this group are wholly composed of soft gelatinous protoplasm or sarcode, the outer layer of which is usually somewhat denser than the inner portion; but both are structureless, except that the inner layer may present a more or less distinct granular appearance. Many of them show a distinct spot or cell, called the nucleus, and some have minute transparent vesicles, which contract and expand alternately, and appear to be of the nature of circulatory or excretory organs. They have no proper alimentary canal, but receive their food into the general mass and digest it in temporary cavities. Their means of locomotion and prehension are soft thread-like or finger-like processes, extended at will from the surface of any part of the body, and known as false feet (pseudopodia). From these processes the whole group has obtained the name of Rhizopods, or rootfooted animals. They may be regarded as constituting the simplest and humblest form of animal life certainly known to us.

The very numerous species of these creatures existing in the waters of the modern world may be arranged under three principal groups. The first and highest includes those which have lobate or finger-like pseudopods, and a well-developed nucleus and pulsating vesicle (Fig. 20, *a*). They are mostly inhabitants of fresh water, and destitute of a hard crust or shell. A second group, including many inhabitants of the sea as well as of fresh waters, has thread-like radiating pseudopodia[4] (Fig. 20 *b*). Some of these form beautiful silicious skeletons. A third group, essentially marine, consists of those with reticulated pseudopodia, and usually destitute of distinct nucleus and pulsating vesicle (Fig. 21). They produce beautiful calcareous skeletons, often very complex, or sometimes are content to cover themselves with a crust of agglutinated grains of sand. It is to this last group that *Eozoon* belongs, and to the highest division of it—that which has the shell perforated with minute pores, often of two kinds. It is curious that just as we have the chambers and pores of *Eozoon* filled with serpentine, so in all geological formations and in the modern seas it is not uncommon to find Foraminifera having their cavities filled with glauconite and other hydrous silicates allied to serpentine.

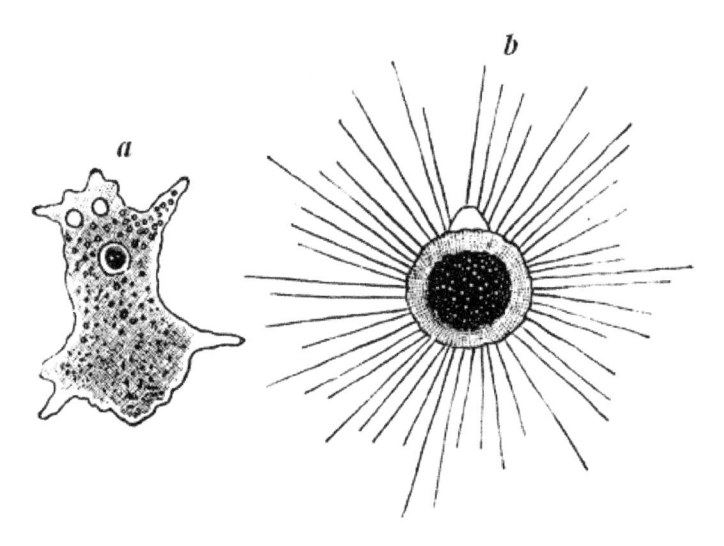

FIG. 20.—*a*, *Amœba*, a fresh-water naked Rhizopod; and *b*, *Actinophrys*, a fresh-water Protozoon of the group Radiolaria, with thread-like pseudopodia.

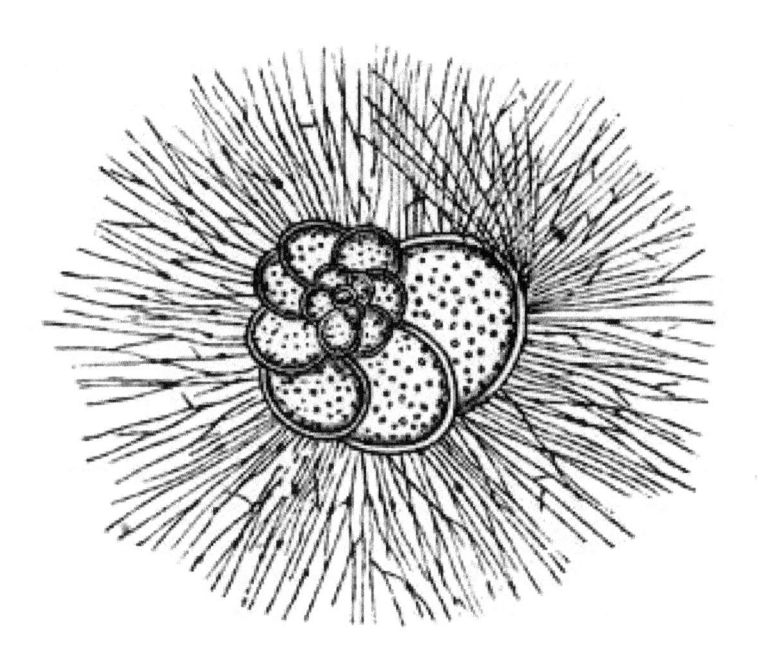

FIG. 21.—*Nonionina*, a modern marine Foraminifer. Showing its chambered shell and netted pseudopodia.—After Carpenter.

If we attempt to trace the Rhizopods onward from the Middle Laurentian, we are met with a great hiatus in the Upper Laurentian. The species *Eozoon Bavaricum* has, however, been found in rocks apparently of Huronian age; but this is the last known appearance of *Eozoon*, properly so-called. In the Cambrian or Siluro-Cambrian, however, we meet with many gigantic Protozoa, more especially those known as *Stromatopora*, *Archæocyathus*, *Receptaculites*, and *Cryptozoon*.

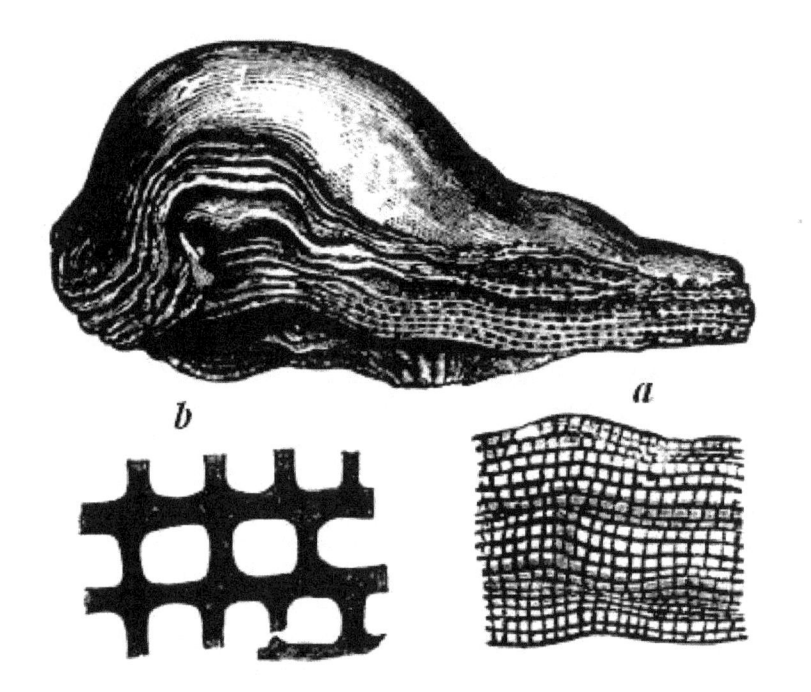

FIG. 22.—*Stromatopora concentrica.*—After Hall.

a, Section of the same, magnified. *b*, Small portion highly magnified, showing laminæ and pillars.

The typical Stromatoporæ, or Layer-corals, consist, like *Eozoon*, of concentric layers, connected by numerous pillars, which are often, though not always, more definite and regular than in the Laurentian fossil. The laminæ are perforated, but more coarsely than in *Eozoon*, and they are often thickened with supplemental deposit which, in some of the forms, presents canals radiating from vertical tubes or bundles of tubes penetrating the mass (Figs. 22, 23). The mode of growth of *Stromatopora* must have closely resembled that of *Eozoon*, and the forms produced are so similar that it is often quite impossible to distinguish them by the naked eye. Like *Eozoon*, they form the substance of important limestones, and single masses are sometimes found as much as three feet in diameter. The Stromatoporæ extend from the Upper Cambrian to the Devonian inclusive. In the Carboniferous they are continued by smaller and more regular organisms of the genus *Loftusia*,[5] and this genus seems to extend without marked change up to the Eocene Tertiary. Recent students of the Stromatoporæ seem disposed to promote them from the province of Protozoa to that of the Hydroids.[6] The reasons for this seem cogent in the case of some of the forms, but in my judgment fail in others, more especially in the older forms. It may ultimately be found that the group as now held includes very different types of structure. In modern times I know of no nearer representative than the animal whose skeleton often adheres in red encrusting patches to our specimens of corals, and which is known as *Polytrema*. In general structure it is not very far from being a very degenerate kind of *Stromatopora*.

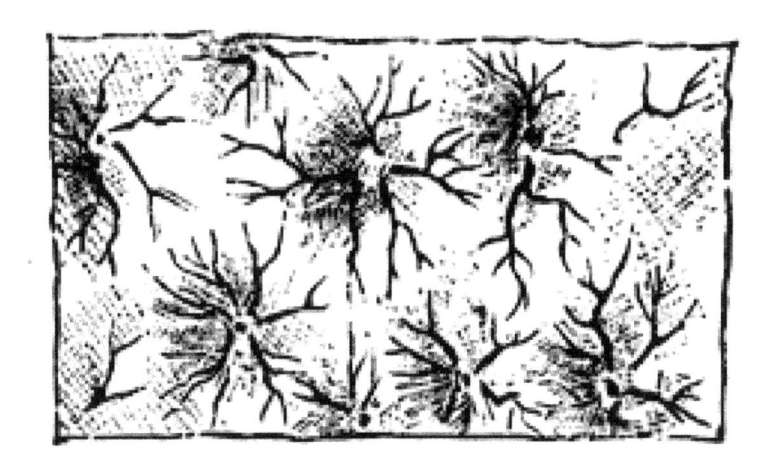

FIG. 23.—*Caunopora planulata*. Showing the radiating canals on a weathered surface. Devonian.—After Hall.

It is curious that in the line of succession above stated, the beautiful tubulated cell-wall of *Eozoon* disappears; and this structure seems, after the Laurentian, to be for ever divorced from the great laminated Protozoans. It reappears in the Carboniferous, in certain smaller organisms of the type of the *Nummulites*, or Money-stone Foraminifers, and is continued in this group of smaller and free animals down to the present time. In the Cretaceous and early Tertiary periods, the Foraminifera of different types have been nearly as great rock-builders as they were in the Laurentian. Some of these later rock-builders, however, have belonged to the lower or imperforate group; others to the higher or Rotaline and Nummuline groups; and, as a whole, they have been individually small, making up in numbers what they lacked in size. Probably the conditions for enabling animals of this type rapidly, and on a large scale, to collect calcareous matter, were more favourable in the Laurentian than they have ever been since.

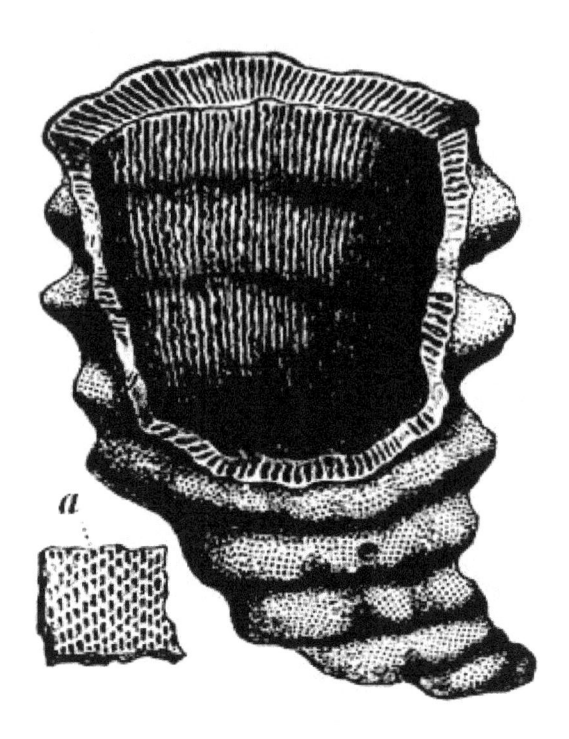

FIG. 24.—*Archæocyathus minganensis.* A Primordial Protozoon.—After Billings.

a, Pores of the inner wall.

In the Siluro-Cambrian age two other forms of gigantic Foraminiferal Protozoans were introduced, widely different from *Eozoon*, and destined apparently not to survive the period in which they appeared. These were *Archæocyathus*, the ancient Cup-corals, and Receptaculites, which may perhaps be called the Sack-corals. Both are quite remote from *Eozoon* in structure, wanting its complexity in the matter of minute tubules, and having greater regularity and complication on the large scale. *Archæocyathus* had the form of a hollow inverted cone with double perforated walls, connected by radiating irregular plates, also perforated (Fig. 24). It has been regarded as a sponge, and some species are certainly accompanied with spicules; but these I have ascertained to be merely accidental, and will be referred to in the next chapter. The true structure of Archæocyathus consists of radiating calcareous plates enclosing chambers connected by pores. *Archæocyathus* came in with the Later Cambrian, and seems to have died out in the Siluro-Cambrian. The only more modern things which at all resemble it are the Foraminifera called *Dactylopora*, which belong to the Tertiary period.

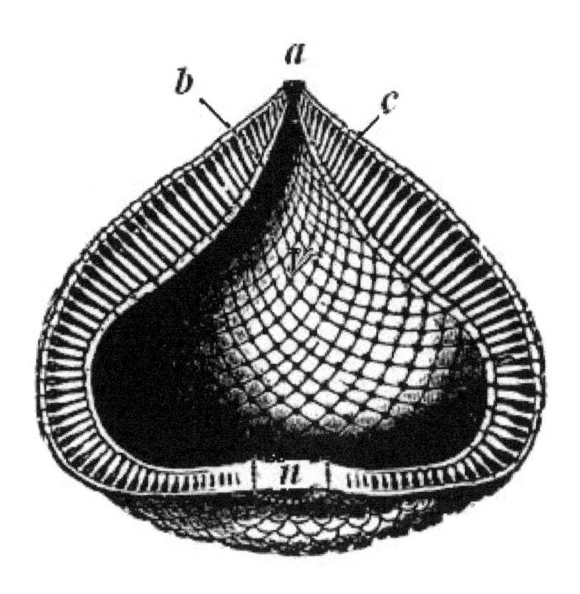

FIG. 25.—Receptaculites. Restored.—After Billings.

a, Aperture. *b*, Inner wall. *c*, Outer wall. *n*, Nucleus, or primary chamber. *v*, Internal cavity.

Receptaculites is a still more complex organism. It has a sack-like form, often attaining a large size, and the double walls are composed of square or rhombic plates, connected with each other by hollow tubes from which proceed canals perforating the plates (Fig. 25). This curious structure is confined to the Siluro-Cambrian, and is so dissimilar from modern forms that its affinities have been subject to grave doubts.

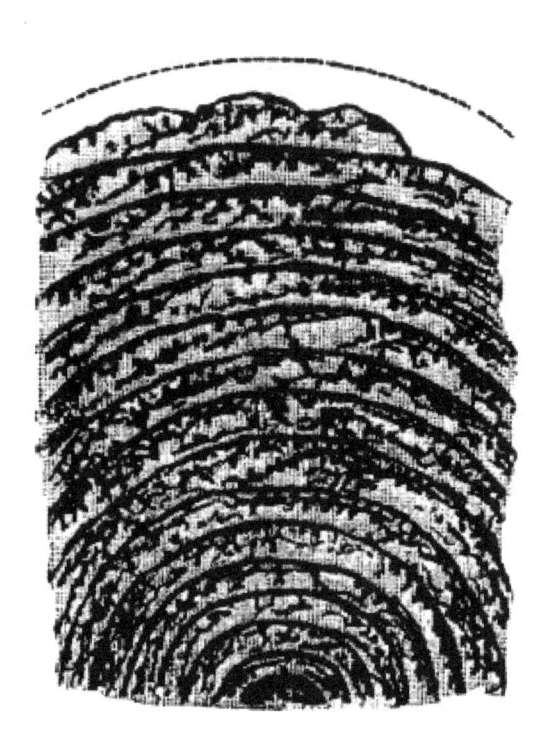

FIG. 26.—Section of *Loftusia Persica*. An Eocene Foraminifer. Magnified five diameters.—After Carpenter and Brady.

We thus have presented to us the remarkable fact that in the Palæozoic age we have no precise representative of *Eozoon*, but instead three divergent types, differing from it and from each other, all apparently specialised to particular uses, all temporary in their duration; while in later times nature seems to have returned nearer to the type of *Eozoon*, though on a smaller scale, and separating some characters conjoined in it. Some portion of this curious result may be due to our ignorance; and it would be interesting to know, what we may know some day, how this type of life was represented in the long interval between the Huronian and the Upper Cambrian, when perhaps there may have been forms that would at least enable us to connect *Eozoon* and *Stromatopora*.

Another link in the chain of being remains to be noticed here. In the Laurentian limestones we meet with numerous minute spherical bodies and groups of spheres with calcareous tubulated tests.[7] These may either be small Foraminiferæ, distinct from *Eozoon*, or may be germs or detached cells from its surface. Similar bodies are found in the lower part of the Siluro-Cambrian, in the Quebec group at Point Levis; and there they are filled with a species of glauconite constituting a sort of greensand rock. Still higher, in the

Carboniferous, there are very numerous species of Foraminifera, presenting forms very similar to those in the modern seas, so that in the smaller shells of this group we seem to have evidence of a continuous series all the way from the Laurentian to the present time. The greater laminated forms co-exist with these up to the Eocene Tertiary. Throughout the whole of geological time—from the formation of the Laurentian limestones to that of the chalky ooze accumulating in the modern ocean—these humble creatures have been among the chief instruments in seizing on the calcareous matter of the waters and depositing it in the form of limestone.

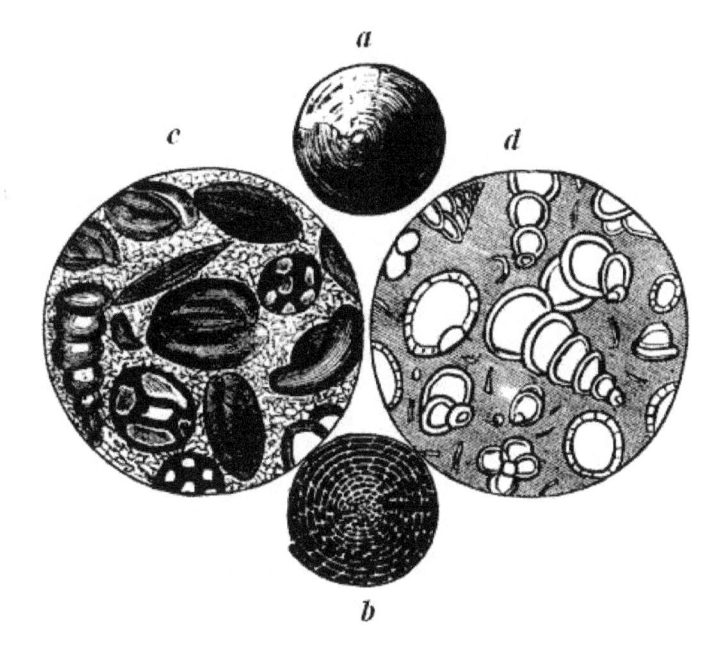

FIG. 27.—Foraminiferal Rock Builders, in the Cretaceous and Eocene.

a, *Nummulites lævigata*—Eocene. *b*, The same, showing chambered interior. *c*, Milioline limestone, magnified—Eocene, Paris. *d*, Hard Chalk, section magnified—Cretaceous.

I have said nothing of the development of higher forms of animal life from *Eozoon*, simply because I know nothing of it. We shall see in the next chapter that these are introduced seemingly in an independent manner. We may be content to trace foraminiferal life along its own line of development, waxing and waning, but ever confined within the same general boundaries, from the Laurentian to the present time. It is likely that if, in any of the ages constituting this vast lapse of time, a dredge had been dropped into the depths of ocean, it would have brought up Foraminifera not essentially different in form and structure. If any one asks to what extent the successive species constituting this almost endless chain may be descendants one of the other, we have no absolutely certain information to give. On the one hand, it is not inconceivable that such forms as *Stromatopora* or *Nummulina* may have descended from *Eozoon*. On the other hand, it is equally conceivable that the same power which produced *Eozoon* at first, whether from dead matter or from some unknown lower form of life, may have repeated the process in later times with modifications. In any case it is probable that the Foraminifera have experienced alternations of expansion and shrinkage, of elevation and decadence, in the lapse of geological time. There were times in which many new forms swarmed into existence, and times in which old forms were becoming extinct without being replaced by others. In so far as the areas of the continents and the adjacent waters are concerned, those periods when the land was subsiding under the ocean must have been their times of prosperity, those in which the crust of the earth shrunk and raised up large areas of land must have been their times of decay. Still this lowest form of animal life has never perished, but has always found abundant place for itself, however pressed by physical change and by the introduction of higher beings.

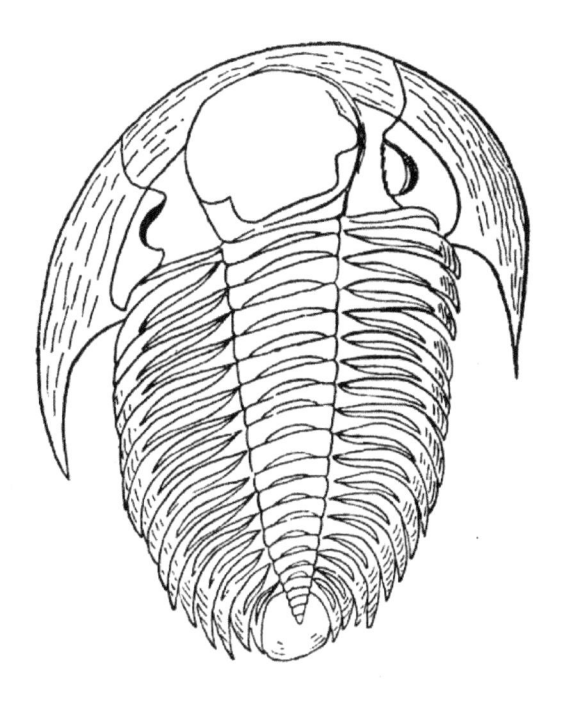

Paradoxides Regina (Matthews). Lower Cambrian of New Brunswick.
1/6th Nat. Size.

CHAPTER III.

THE AGE OF INVERTEBRATES OF THE SEA.

IF the middle portion of the Laurentian age was really a time of exuberant and abounding life, either this met with strange reverses in succeeding periods, or the conditions of preservation have been such as to prevent us from tracing its onward history. Certain it is, that according to present appearances we have a new beginning in the Cambrian, which introduces the great Palæozoic age, and few links of connection are known between this and the previous Eozoic.

At the beginning of the Palæozoic we have reason to believe that our continents were slowly subsiding under the sea, after a period of general continental elevation which was consequent on the crumbling of the earth's crust at the close of the Eozoic; and on the new sea-bottoms formed by this subsidence came in, slowly at first, but in ever-increasing swarms, the abundant and varied life of the early Palæozoic.

In the oldest portion of the Cambrian series in Wales, Hicks has catalogued species of no less than seventeen genera, embracing Crustaceans, the representatives of our crabs and lobsters, bivalve and univalve shell-fishes of different types, worms, sea-stars, zoophytes, and sponges. If we could have walked on the shores of the old Cambrian sea, or cast our dredge or trawl into its depths, we should have found representatives of most of the humbler forms of sea life still extant, though of specific forms strange to us. Perhaps the nearest approach to such experience which we can make is to examine the group of Cambrian animals delineated in Fig. 28, and to notice, under the guidance of the geologist above named, the sections seen at St. David's, in South Wales.

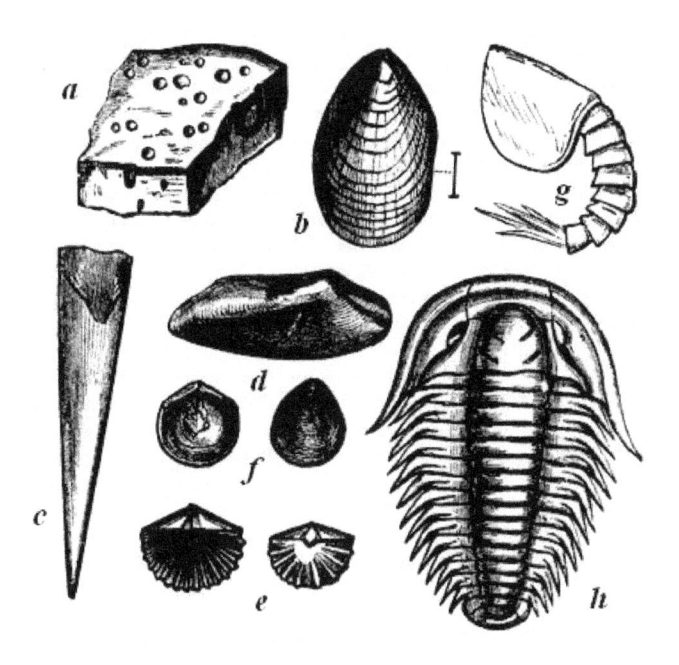

FIG. 28.—Group of Cambrian Animals (from Nicholson).

a, Arenicolites didymus, worm tubes. *b, Lingulella ferruginea. c, Theca Davidii. d, Modiolopsis solvensis. e, Orthis Hicksii. f, Obolella sagittalis. g, Hymenocaris vermicauda. h,* Trilobite, *Olenus micrurus.*

Here we find a nucleus of ancient rocks supposed to be Laurentian, though in mineral character more nearly akin to the Huronian, but which have hitherto afforded no trace of fossils. Resting unconformably on these is a series of partially altered rocks, regarded as Lower Cambrian, and also destitute of organic remains. These have a thickness of almost 1,000 feet, and they are succeeded by 3,000 feet more of similar rocks, still classed as Lower Cambrian, but which have afforded fossils. The lowest bed which contains indications of life is a red shale, perhaps a deep-sea bed, and possibly itself partly of organic origin, by that strange process of decomposition or dissolution of foraminiferal ooze and volcanic fragments, going on in the depths of the modern ocean, and described by Dr. Wyville Thomson as occurring over large areas in the South Pacific. The species are two *Lingulellæ*, a *Discina* and a *Leperditia*. Supposing these to be all, it is remarkable that we have no Protozoa or Corals or Echinoderms, and that the types of Brachiopods and Crustaceans are of comparatively modern affinities. Passing upward through another 1,000 feet of barren sandstone, we reach a zone in which no less than five genera of Trilobites are found, along with Pteropods and a sponge. Thus it is that life comes in at the base of the Cambrian in Wales, and it may be regarded as a fair specimen of the facts as they appear in the earlier fossiliferous beds succeeding the Laurentian. Taking the first of these groups of fossils, we may recognise in the *Leperditia* a two-valved Crustacean closely allied to forms still living in the seas and fresh waters. The Lingulellæ, whether we regard them as molluscoids, or, with Professor Morse, as singularly specialised worms, represent a peculiar and distinct type, handed down, through all the vicissitudes of the geological ages, to the present day. The Pteropods and the sponge are very similar to forms now living. The Trilobites are an extinct group, but closely allied to some modern Crustaceans. Had the primordial life begun with species altogether inscrutable and unexampled in succeeding ages, this would no doubt have been mysterious; but next to this is the mystery of the oldest forms of life being also among the newest. Whatever the origin of these creatures, they represent families which have endured till now in the struggle for existence without either elevation or degradation. Yet, though thus vast in their duration, they seem to have swarmed in together and in great numbers, in the Cambrian, without any previous preparation. From the Cambrian onward, throughout the whole Palæozoic, there is no decided break in the continuity of marine life; and already in the Silurian period the sea was tenanted with all the forms of invertebrate life it yet presents, and these in a teeming abundance not surpassed in any succeeding age. Let us now, in accordance with our plan, select some of these ancient inhabitants of the waters and trace their subsequent history.

Remains of sea-weeds are undoubtedly present in the Cambrian rocks. One of the lowest beds in Sweden has been named from their abundance the Fucoidal Sandstone; and wherever fossiliferous Cambrian rocks occur, some traces, more or less obscure, of these plants may be found. Nearly all that we can say of them, however, is, that, in so far as their remains give any information, they are very like the plants of the same group that now abound in our seas. In the fucoidal sandstone of Sweden certain striated or ribbed bodies have been found, which have even been regarded as land plants;[8] but they seem rather to be trails or marks left by sea-weeds dragged by currents over a muddy bottom. The plants of the sea thus precede those of the land, and they begin on the same level as to structure that they have since maintained. I agree with Nathorst, however, in holding that the Bilobites and many other forms believed by some to be sea-weeds, are really trails and tracks of animals.[9]

The Foraminifera of the Palæozoic we have noticed in the last chapter; but we now find a new type of Protozoan—that of the Sponge. Sponges as they exist at present may be defined to be composite animals, made up of a great number of one-celled or gelatinous zoids, provided with vibrating threads or cilia, and so arranged that currents of water are driven through passages or canals in the mass, by the action of the cilia, bringing food and aerated water for respiration. To support these soft sarcodic sponge-masses, they secrete fibres of horny matter and needles (spiculæ) of flint or of limestone, forming complicated fibrous and spicular skeletons, often of great beauty. They abound in all seas, and some species are found in fresh waters.

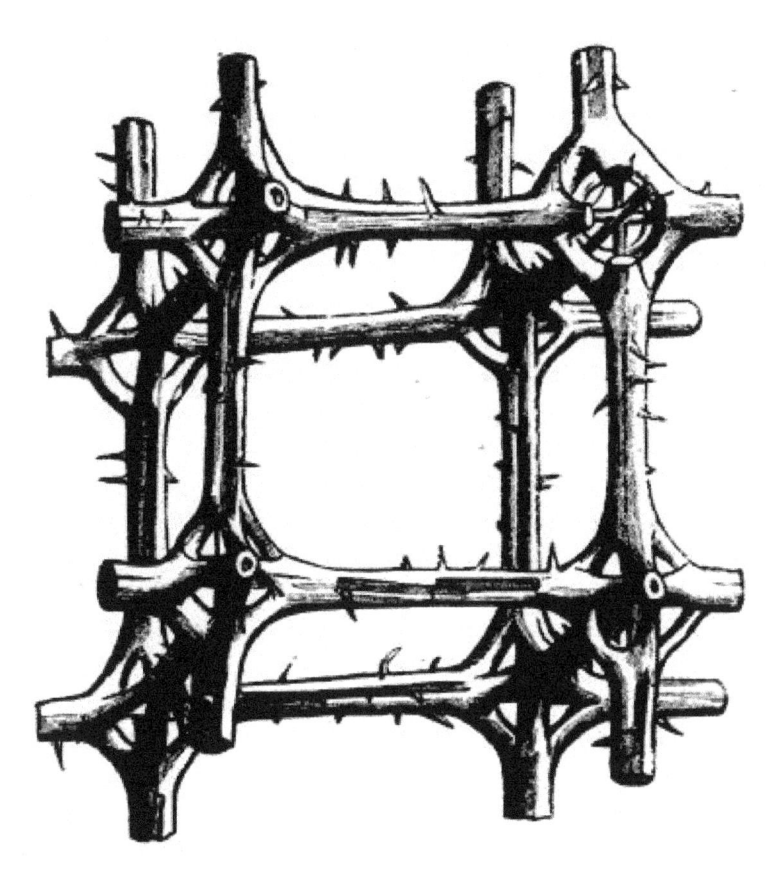

FIG. 29.—Portion of skeleton of Hexactinellid Sponge (*Cœloptychium*). Magnified. After Zittel.

With the exception of a very few species destitute of skeleton, and which we cannot expect to find in a fossil state, the sponges may be roughly divided into three groups: 1, those with corneous or horny skeleton, like our common washing sponges; 2, those with skeletons composed of silicious needles of various forms and arrangement; 3, those with calcareous spicules. Of these, the second or silicious group has precedence in point of time, beginning in the Early Cambrian, and continuing to the present. Two of its subdivisions are especially interesting in their range. The first is that of the Lattice-sponges (*Hexactinellidæ*), in which the spicules have six rays placed at right angles, and are attached to each other by their points, so as to form a very regular network (Fig. 29). The second is that of the Stone-sponges (*Lithistidæ*), in which the spicules are four-rayed or irregular, and are united by the branching root-like ends of the rays. The most beautiful of all sponges, the Venus Flower-basket (*Euplectella*), is a modern Hexactinellid, and the wonderful weaving of its spicules is as marvellous a triumph of constructive skill as its general form is graceful. The Lithistids are less beautiful, but are the densest and most compact of sponges, and are represented by several species in the modern seas. Both of these types go back to the Early Cambrian, and have continued side by side to the present day, as representatives of two distinct geometrical methods for the construction of a spicular skeleton.

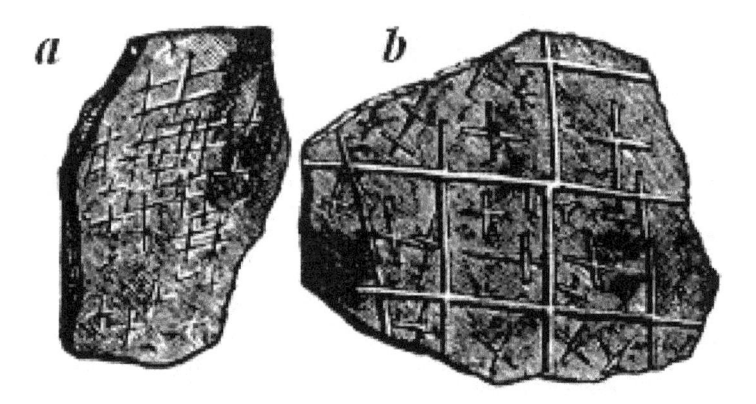

FIG. 30.—*Protospongia fenestrata* (Salter). Menevian group.

a, Fragment showing the spicules partially displaced. *b*, Portion enlarged.

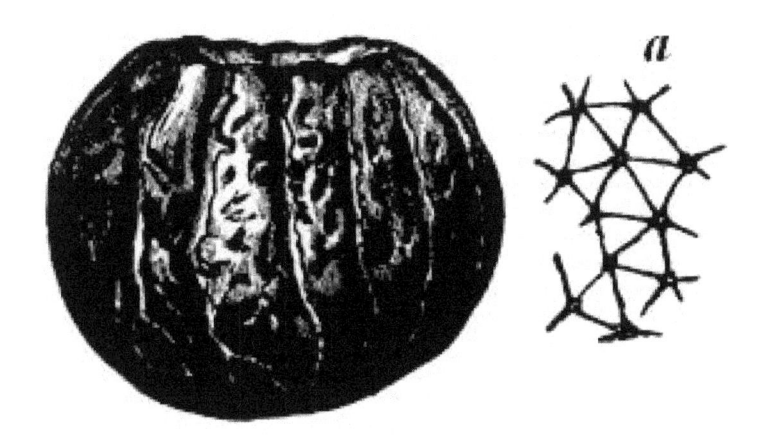

FIG. 31.—*Astylospongia præmorsa* (Roemer). Niagara group.—After Hall.
a, Spicules magnified.

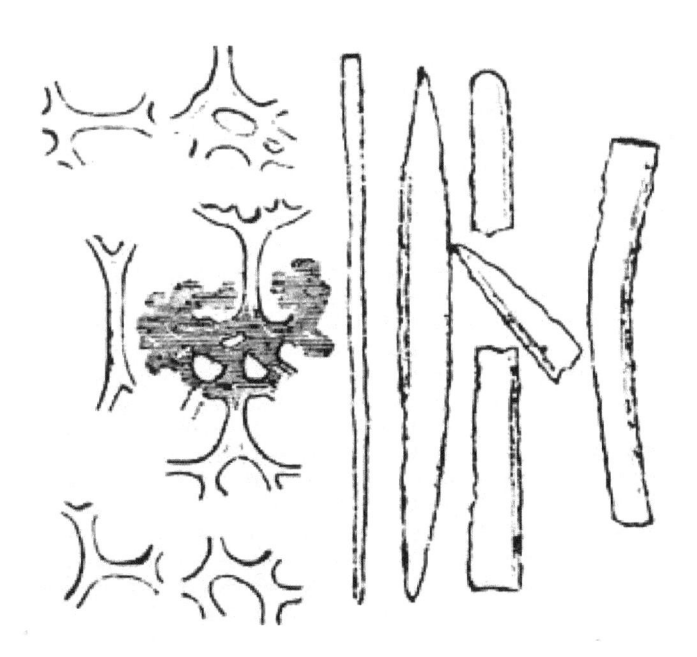

FIG. 32.—Spicules of Lithistid sponge (*Trichospongia* of Billings). From the Cambrian of Labrador.

Many years ago the keen eye of the late lamented Salter detected in a stain on the surface of a slab of Cambrian slate the remains of a sponge; and minute examination showed that its spicules crossed each other, and formed lattice-work on the hexactinellid plan. Salter boldly named it *Protospongia* (the first sponge), and it is still the earliest that we know (Fig. 30). Thus the type whose skeleton is the most perfect in a mechanical point of view takes the lead. It is continued in the Silurian in many curious forms, of which the stalkless sponges (*Astylospongia*) are the most common (Fig. 31). It perhaps attains its maximum in the Cretaceous, from which the beautiful example in Fig. 29 is taken, and it still flourishes, giving us the most beautiful of all recent forms. Before the close of the Cambrian there were other sponges of the Lithistid type. Fig. 32 represents a group of spicules from the Calciferous (Lowest Silurian or Upper Cambrian) of Mingan,[10] and which probably belong to a large Lithistid sponge of that early time. The Lithistids have been recognised in the Upper Silurian and Carboniferous, and continuing upward to the Cretaceous, there become vastly numerous, while their modern representatives are by no means few. The silicious sponges with simple spicules appear to have existed as far back as the Siluro-Cambrian, and there is believed to be almost as early evidence of horny or corneous sponges. The calcareous sponges have been recognised as far back as the Silurian.[11] Thus from the close of the Palæozoic all the types of sponges seem to have existed side by side; and in the Cretaceous period, when such large areas of our continents were deeply submerged, they attained a wonderful development, perhaps not equalled in any other era of the earth's history.

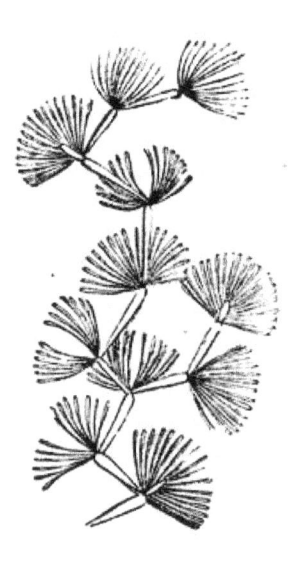

FIG. 33.—*Oldhamia antiqua* (Forbes).

FIG. 34.—*Dictyonema sociale*. Enlarged. *Lingula* flags.—After Salter.

Sponges may be regarded as the highest or most complex of the Protozoa or the lowest of the Coelenterates. We have no links wherewith to connect them with the lower Protozoa of the Eozoic period; and through their long history, though very numerous in genera and species, they show no closer relationship with the Foraminifera below, and the Corals above, than do their successors in the modern seas. They thus stand very much apart; and modern studies of their development and minute structures do not seem to remove them from this isolation. Though we are treating here of inhabitants of the sea, it may be proper to mention that Geinitz has described two species from the Permian which he believed to be early precursors of the Spongillæ, or fresh-water sponges; but more recently he seems to regard them as probably Algæ. Young has, however, recently found true spicules of *Spongilla* in the Purbeck beds.[12]

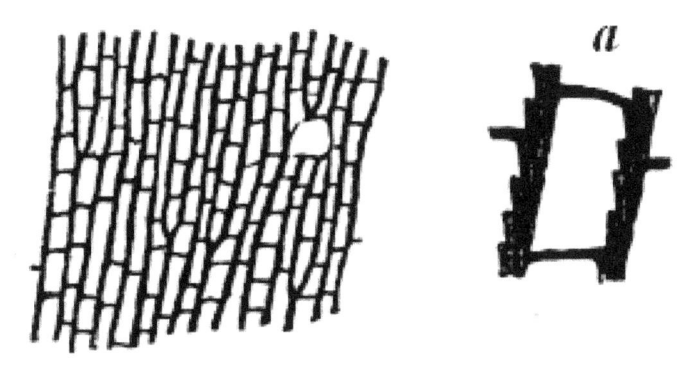

FIG. 35.—*Dictyonema Websteri* (Dn). Niagara formation.

a, Enlarged portion (*Acadian Geology*).

FIG. 36.—Group of modern Hydroids allied to Graptolites. Magnified, and natural size.

a, Sertularia. b, Tubularia. c, Campanularia.

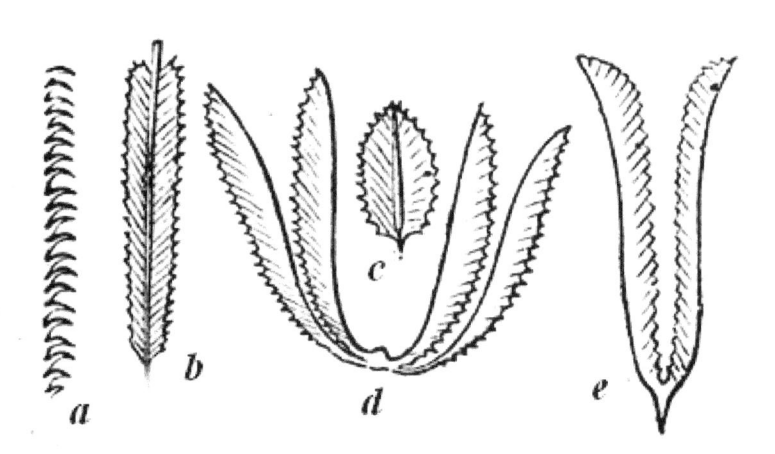

FIG. 37.—Silurian Graptolitidæ.

- 50 -

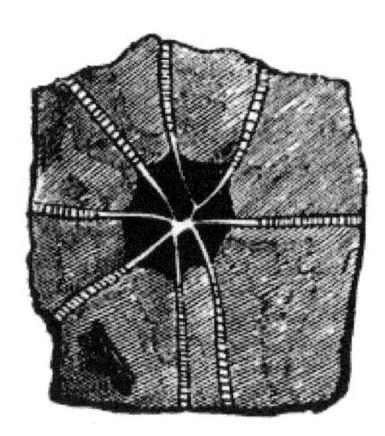

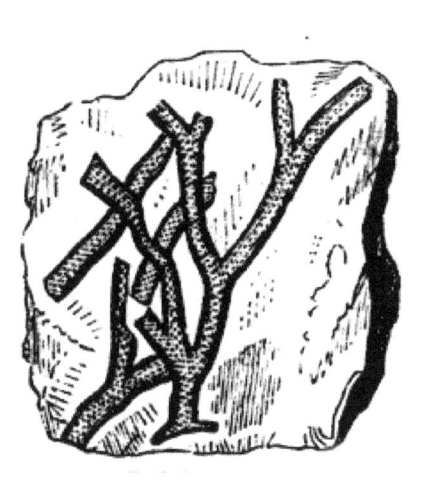

a, *Graptolithus*. *b*, *Diplograpsus*. *c*, *Phyllograpsus*. *d*, *Tetragrapsus*. *e*, *Didymograpsus*.

FIG. 38.—Central portion of Graptolite, with membrane, or float (*Dichograpsus octobrachiatus*, Hall).

FIG. 39—*Ptilodictya acuta* (Hall). Bryozoan. Siluro-Cambrian.

A stage higher than the sponges are those little polyp-like animals with sac-like bodies and radiating arms or tentacles, which form minute horny or calcareous cells, and bud out into branching communities, looking to untrained eyes like delicate sea-weeds—the sea-firs and sea-mosses of our coasts (Fig. 36). These belong to a very old group, for in the oldest Cambrian we have a form referred to this type (Fig. 33), and in the Upper Cambrian another still more decided example (Fig. 34).[13] This style of life, once introduced, must have increased in variety and extended itself with amazing rapidity, for in the Siluro-Cambrian age we find it already as characteristic as in our modern seas, and so abundant that vast thicknesses of shale are filled and blackened with the *débris* of forms allied to the sea-firs, and masses of limestone largely made up of the more calcareous forms of the sea-mosses. As examples of the former we may take the *Graptolites*, so named from their resemblance to lines of writing, and of which several forms are represented in Fig. 37. The little teeth on the sides of these were cells, inhabited probably by polyps, like those represented in the modern *Sertularia* in Fig. 36. Some of them were probably attached to the bottom. In others the branches radiated from a central film which may have been a hollow vesicle or float, enabling them to live at the surface of the water (Fig. 38). These Graptolites are specially characteristic of the Upper Cambrian and Lower Silurian. The netted ones (*Dictyonema*), as may be seen from Figs. 34 and 35, came in before the close of the Cambrian, and continue unchanged to the Silurian, where they disappear. The branching forms, seen in Fig. 37, have scarcely so great a range. They thus form most certain marks of the period to which they belong, and being oceanic and probably floaters, they diffused themselves so rapidly that they appear to indicate the same geological time in countries so widely separated as Europe, North America, and Australia. It is curious, too, that while the Graptolites thus mark a definite geological time, and seem to disappear abruptly and without apparent cause, they are the first link in the long chain of the Hydroids, which, though under different family forms, continue to this day, apparently neither better nor worse than their perished Palæozoic relatives. There is a group of little Stony Corals (*Monticuliporidæ*), which were possibly also the cells of Hydroids, that have a similar history. They are the only known Corals that date so far back as the Upper Cambrian; and they continue under very similar forms all through the Palæozic, and are represented by the millepore corals of the present day. Fig. 40 represents a form found at the base of the Siluro-Cambrian, and Fig. 41 shows forms characteristic of the Carboniferous Limestone.

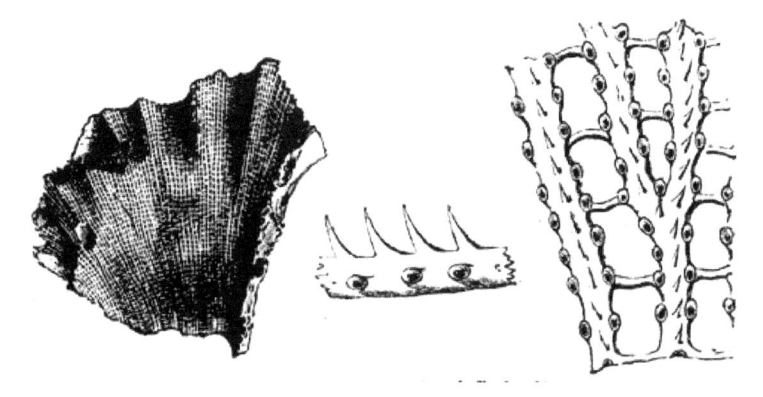

FIG. 39*a*.—*Fenestella Lyelli* (Dawson). A Carboniferous Bryozoan.

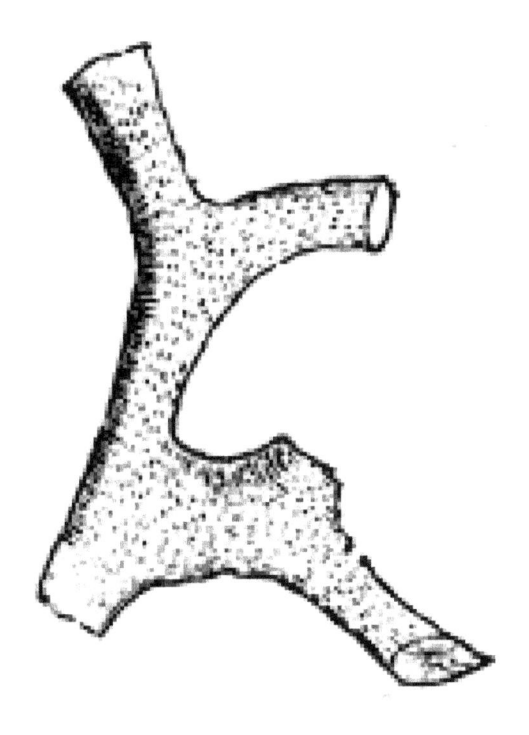

If we turn now to the sea-mosses (Bryozoa), we have a group of minute polyp-like animals

FIG. 40.—*Chaetetes fibrosa*. A tubulate coral with microscopic cells. Siluro-Cambrian. inhabiting cells not unlike those of the Hydroids, and which form plant-like aggregates. But the animals themselves are so different in structure that they are considered to be nearer allies of the bivalve shell-fishes than of the Corals. They are, in short, so different, that the most ardent evolutionist would scarcely hold a community of origin between them and such creatures as the Graptolites and Millepores, though an ordinary observer might readily confound the one with the other. These animals appear at the beginning of the Siluro-Cambrian, and such forms as that represented in Fig. 39, very closely allied to some now living, are large constituents of some of the limestones of that period. Other forms, like that represented in Fig. 39*a*, are very characteristic of the Carboniferous. These animals, individually small, though complicated in structure and branching into communities, scarcely ever of any great magnitude, humble creatures which have never played any great part in the world, have, nevertheless, been so persistent that, though specific and generic forms have been changed, the group may be said to be in the modern seas exactly what it was in those of the early Palæozoic, nor can it be affirmed to have originated in anything different, or to have produced anything.

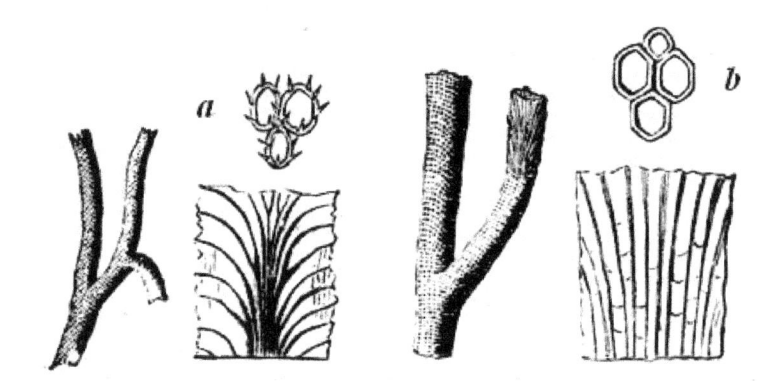

FIG. 41.—*a, Stenopora exilis* (Dawson). *b, Chaetetes tumidus* (Edwards and Haine). Carboniferous.

The true Stony Corals (*Anthozoa*) are as yet unknown in the Cambrian. They entered on the stage in immense abundance in the Siluro-Cambrian, where considerable limestones are largely composed of their remains, mixed, however, and sometimes overpowered with those of Bryozoa and Hydroids. An ordinary coral, such as those of which coral reefs are built—the red coral, used for ornament is not quite similar—is the skeleton of an animal constructed on the plan of a sea anemone; with a central stomach surrounded by radiating chambers, and having above a crown of tentacles. The stony coral surrounds and protects the soft body of the animal, and may either be a single cell, for one animal, or an aggregation of such cells, constituting a rounded or branching mass. The modern star coral, represented in Fig. 42, is an instance of the latter condition. It shows nineteen or twenty animals, each with a central mouth and fringe of short tentacles, aggregated together, and two of them showing the spontaneous division by which the number of animals in the mass is progressively increased. The living coral shows only the soft animals and the animal matter connecting them; but if dead there would be a white stony mass with a star-like cell or depression corresponding to each animal.

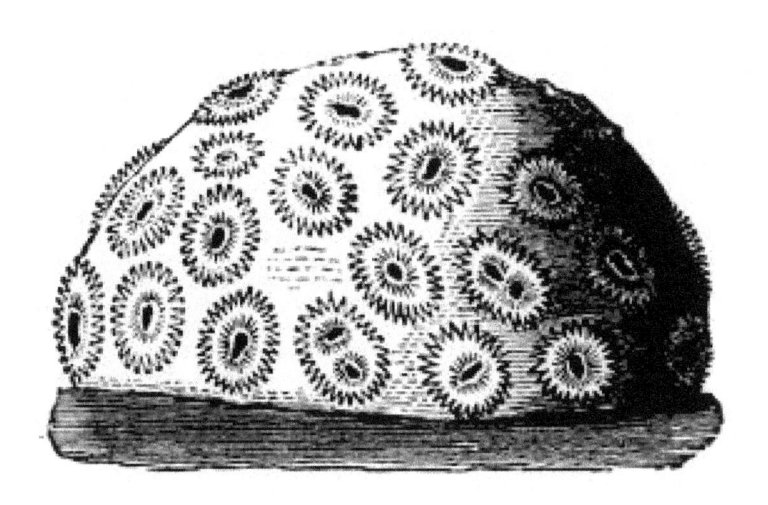

FIG. 42.—Living Anthozoan Coral (*Astræa*).

In their general plan, the oldest Corals were precisely of this character, but they presented some differences in detail, which have caused them to be divided into two groups, which are eminently characteristic of the Palæozoic age—the tabulate or floored corals, and the rugose or wrinkled corals. In the former (Fig. 43) the cells are usually small and thin-walled, often hexagonal, like a honeycomb, and are floored across at intervals with tabulæ or horizontal plates. A few modern corals present a similar arrangement,[14] but this kind of structure was far more prevalent in the Palæozoic. In the second type the animals are usually larger and often solitary, the cell has strongly marked radiating plates, while the horizontal floors are absent or subordinate, and there is usually a thick external rind or outer coat (Figs. 44, 45). In general plan, these rugose corals closely resemble those of our modern reefs; but they differ in their details of structure, and only a very few modern forms from the deep sea are regarded as actual modern representatives.[15] One curious point of difference is that their radiating laminæ begin with four, and increase by multiples of that number, while in modern corals the numbers are six and multiples of six; a change of mathematical relation not easily accounted for, and which assimilates them to Hydroids on the one hand, and to a higher group, the Alcyonids, on the other, both of which prefer four and eight to six, or have had these numbers chosen for them. In the Mesozoic period the tabulate and rugose corals were replaced by others, the porous and solid corals of the modern seas; but, in so far as we know, the animals producing these, though differing in some details, were neither more nor less elevated than their predecessors, and they took up precisely the same rôle as reef-builders in the sea, though with probably more tendency to the accumulation of great masses of coral limestone in particular spots.

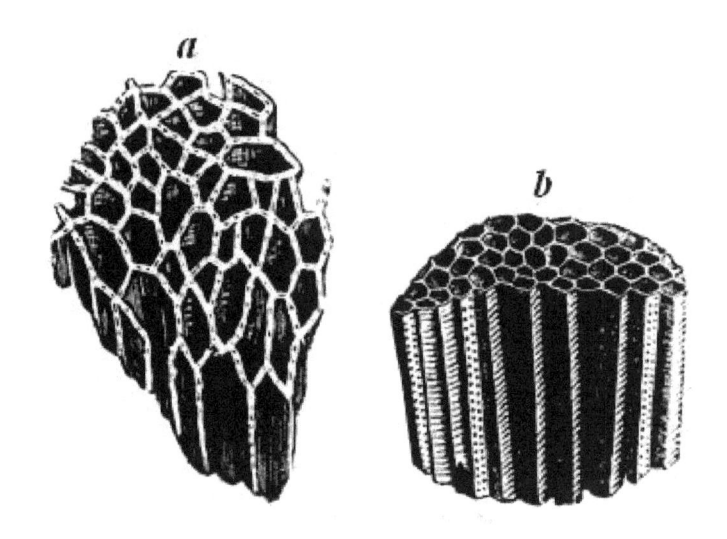

FIG. 43.—Tabulate Corals.
a, *Halisites*, and *b*, *Favosites*. Upper Silurian.

FIG. 44.—Rugose Coral (*Heliophyllum Halli*). Devonian.

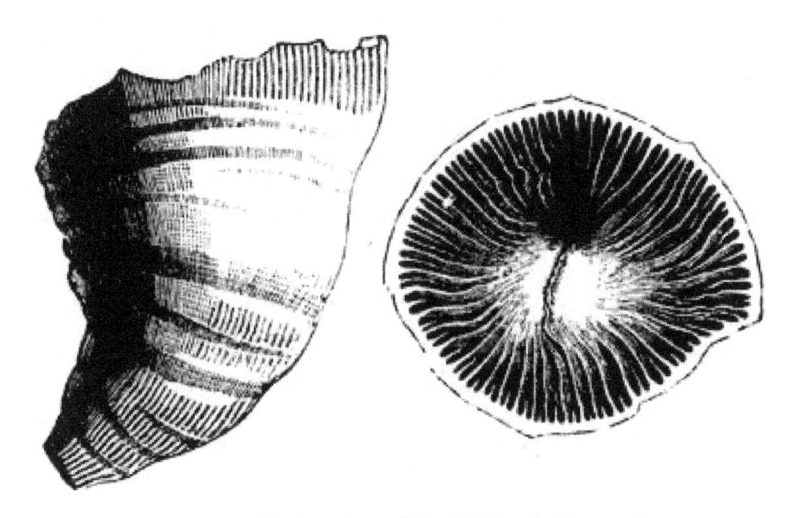

FIG. 44a.—*Zaphrentis prolifica* (Billings). Devonian.

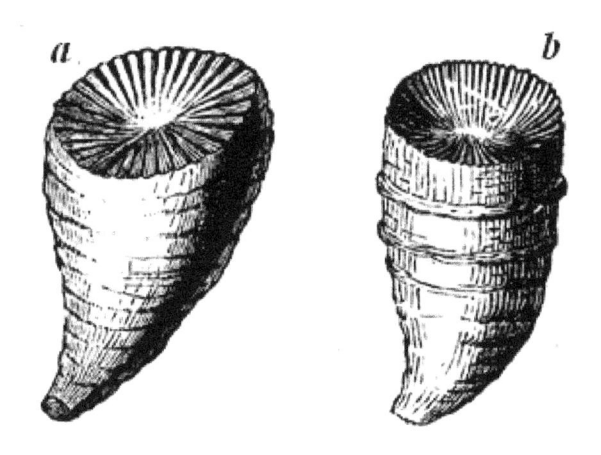

FIG. 45.—Rugose Corals.

a, *Zaphrentis Minas* (Dn.), and b, *Cyathophyllum Billingsi* (Dn.). Carboniferous.

Leaving the corals, we may turn to the sea-stars and seaurchins. These merely put in an appearance in the Early Cambrian, but become vastly multiplied in the Silurian, where the stalked feather stars (Crinoids) (Fig. 46) seem to have covered great areas of sea-bottom, and multiplied so rapidly that thick sheets of limestone are largely made up of the fragments of their skeletons. The ordinary star-fishes appear first in the Silurian (Fig. 47). The sea-urchins begin in the Upper Silurian, the early species having numerous and loosely attached plates, like some of those now found in the deep sea[16] (Fig. 48).

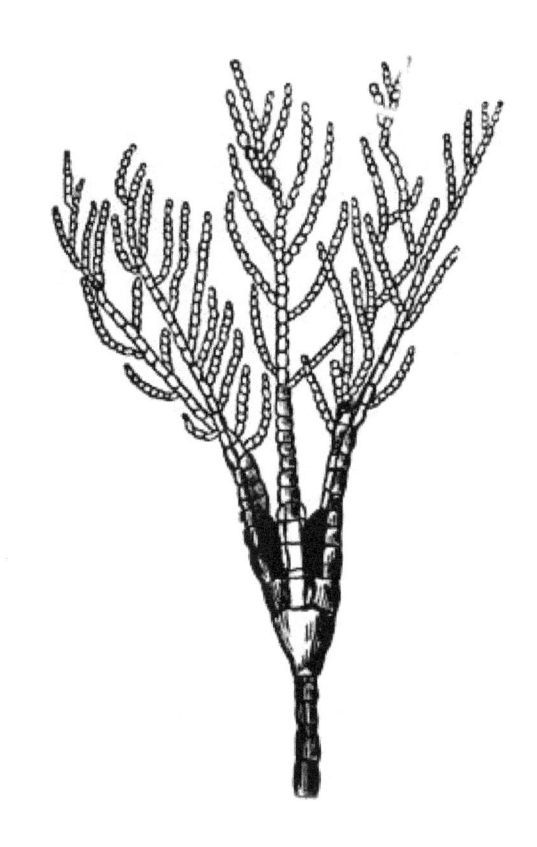

Fig. 46.—Modern Crinoid (*Rhisocrinus Lofotensis*).—After Sars.

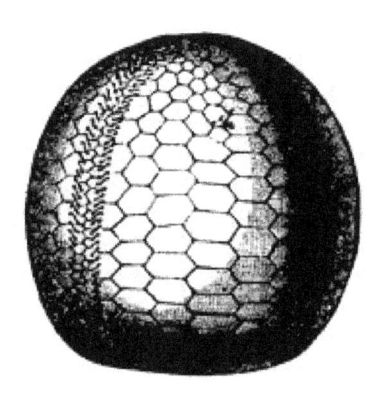

IG. 47.—*Palæaster Niagarensis* (Hall). Fig. 48.—*Palæchinus ellipticus* (McCoy).
One of the oldest star fishes.   One of the oldest types of sea-urchins.

The most curious history in this group is that of the feather-stars. In the Early Cambrian they are represented by a few species known to us only in fragments, and these belong to a humble group (Cystideans) resembling the larval or immature condition of the higher Crinoids. Fig. 49 shows one of these animals of somewhat later age. They have few or rudimentary arms and short stalks, and want the beautiful radial symmetry of the typical star-fishes. In the Silurian these creatures are reinforced by a vast number of beautiful and perfect feather-stars (Figs. 50, 51). These continue to increase in number and beauty, and apparently culminate in the Mesozoic, where gigantic forms exist, some of them probably having more complicated skeletons, in so far as number of distinct parts is concerned, than any other animals. Buckland has calculated that in a crinoid similar to that in Fig. 52 there are no less than 150,000 little bones, and 300,000 contractile bundles of fibres to move them. In the modern seas the feather-stars have somewhat dwindled both in numbers and complexity, and are mostly confined to the depths of the ocean. On the other hand, the various types of ordinary star-fishes and sea-urchins have increased in number and importance. We thus find in this group a certain advance and improvement from the Cystideans of the Early Palæozoic to the sea-urchins and their allies. This advance is not, however, along one line for the Cystideans continue unimproved to the end. The Crinoids culminate in the Mesozoic, and are not known to give origin to anything higher. The star-fishes and sea-urchins commence independently, before the culmination of the Crinoids, and, though greatly increased in number and variety, still adhere very closely to their original types.

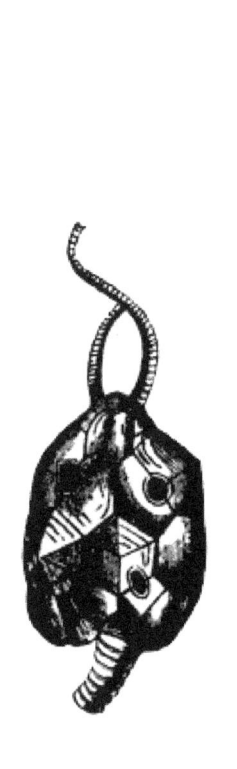

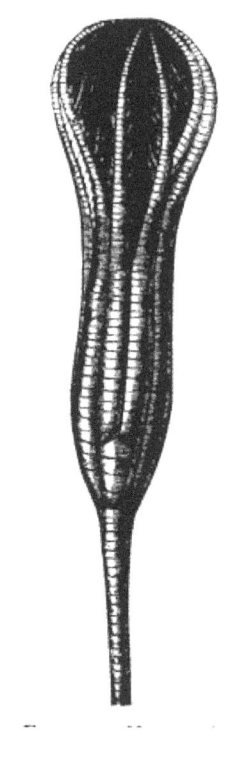

FIG. 49.—*Pleurocystites squamosus*. Siluro-Cambrian. After Billings.

FIG. 50.—*Heterocrinus simplex* (Meek). One of the least complex crinoids of that period. Siluro-Cambrian.

FIG. 51.—Body of *Glyptocrinus*. Siluro-Cambrian.

The great sub-kingdom of the Mollusca, including the bivalve and univalve shell-fishes, makes its first appearance in the Cambrian, where its earliest representatives belong to a group, the Arm-bearers or Lamp shells (Brachiopods), held by some to be allied to worms as much as to mollusks. The oldest of all these shells are allies of the modern *Lingulæ* (Fig. 54), some of the earliest of which are shown in Fig. 55. The modern *Lingula* is protected by a delicate two-valved shell, composed, unlike that of most other mollusks, of phosphate of lime or bone earth. It lives on sand-banks, attached by its long flexible stalk, which it buries like a root in the bottom. Its food consists of microscopic organisms, drifted to its mouth by cilia placed on two arm-like processes, from which the group derives its name. In the modern world about one hundred species of Brachiopods are known, belonging to about twenty genera, some of which differ considerably from the Lingulæ. The genus *Terebratula*, represented at Fig. 56, is one of the most common modern as well as fossil forms, and has the valves unequal, with a round opening in one of them for the stalk, which is attached to some hard object, and there is an internal shelly loop for supporting the arms.

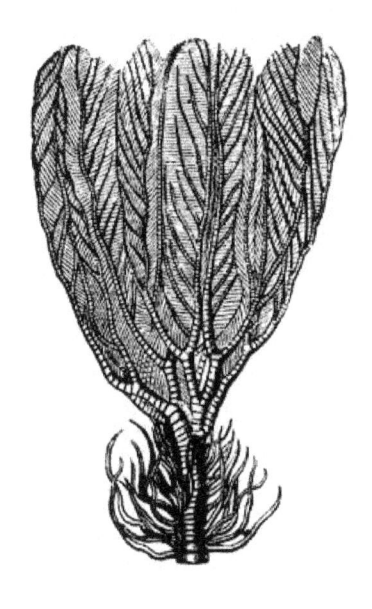

FIG. 52.—*Extracrinus Briareus*. Reduced. Jurassic.

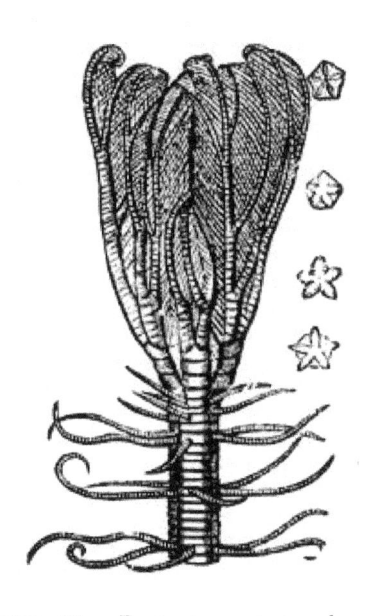

FIG. 53.—*Pentacrinus caput-medusæ*. Reduced. Modern.

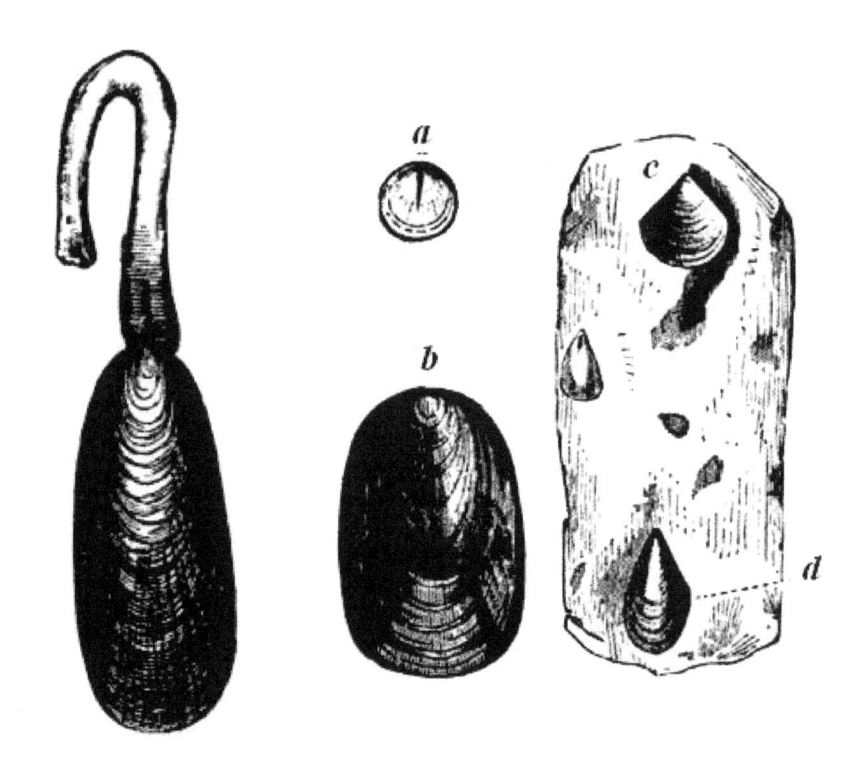

FIG. 54.—*Lingula anatina*.
With flexible muscular stalk. Modern.

FIG. 55.—Cambrian and Silurian Lingulæ.

a, *Lingulella Matthewi* (Hartt). Acadian group.
b, *Lingula quadrata* (Hall). Siluro-Cambrian.
c, *Lingulella prima* (Hall). Potsdam. d, *Lingulella antiqua* (Hall). Potsdam.

These curious, and in the modern seas, exceptional shells, were dominant in the Palæozoic period. Upwards of three thousand fossil species are known, of which a large proportion belong to the Cambrian and Silurian, nine genera appearing in the Cambrian, and no less than fifty-two in the Silurian. The history of these creatures is very remarkable. The Lingulæ, which are the first to appear, continue unchanged and with the same phosphatic shells to the present day. Morse, who has carefully studied an American species, remarks in illustration of this, that it is exceedingly tenacious of life, bearing much change of depth, temperature, etc., without being destroyed. The genus *Discina*, which is nearly as old, also continues throughout geological time. The genus *Orthis* (Fig. 57), which appears at the same time with the last, becomes vastly abundant in Silurian times, but dies out altogether before the end of the Palæozoic. *Rhynchonella* (Fig. 58), which comes in a little later, near the beginning of the Siluro-Cambrian, continues to this day. *Spirifer* and *Productus* (Figs. 59 and 60) appear later, and die out at the close of the Palæozoic. So strange and inscrutable are the fortunes of these animals, which on the whole have lost in the battle of life, that their place in nature is vastly less important than it was. It has been suggested that if any group of creatures could throw light upon the theory of descent with modification, it would be these; but Davidson, who has perhaps studied them more thoroughly than any other naturalist, found them as silent on the subject as the sponges or the corals. In a series of papers published in the *Geological Magazine*, a short time before his death, he remarked as follows:

FIG. 56.—*Terebratula sacculus* (Martin). Carboniferous.

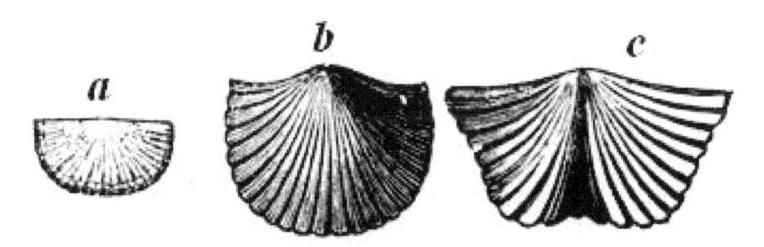

FIG. 57.—Brachiopods; genus *Orthis*.

a, *O. Billingsi* (Hartt). Lower Cambrian. *b*, *O. pectinella* (Hall). Siluro-Cambrian. *c*, *O. lynx* (Eichwald). Siluro-Cambrian.

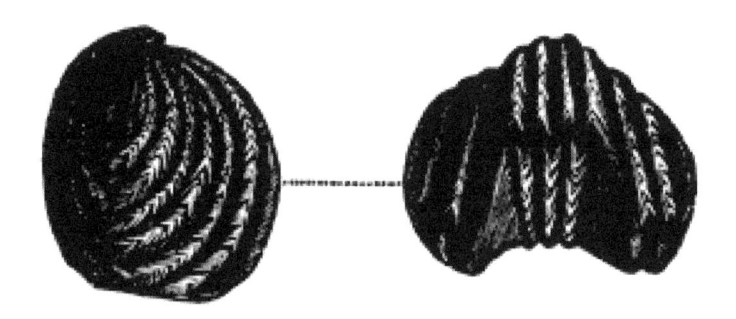

FIG. 58.—*Rhynchonella increbrescens* (Hall). Siluro-Cambrian.

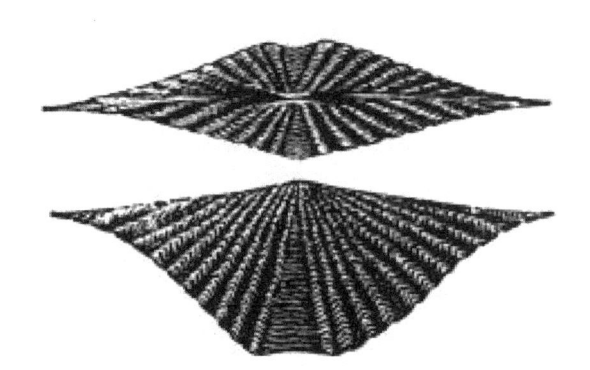

FIG. 59.—*Spirifer mucronatus* (Conrad). Devonian.

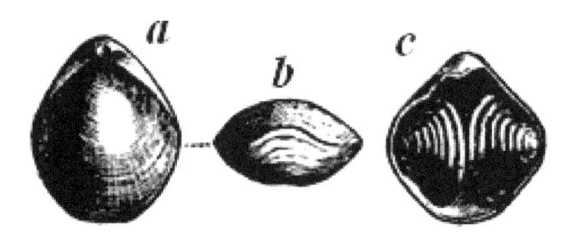

FIG. 59*a*.—*Athyris subtilita* (Hall). Carboniferous.

a, *b*, Exteriors. *c*, Interior, showing spirals.

"We find that the large number of genera made their first appearance during the Palæozoic periods, and since they have been decreasing in number to the present period. We will leave out of question the species, for they vary so little that it is often very difficult to trace really good distinctive characters between them; it is different with the genera, as they are, or should be, founded on much greater and more permanent distinctions. Thus, for example, the family *Spiriferidæ* includes genera which are all characterised by a calcified spiral lamina for the support of the brachial appendages; and, however varied these may be, they always retain the distinctive characters of the group from their first appearance to their extinction. The Brachiopodist labours under the difficulties of not being able to determine what are the simplest, or which are the highest families into which either of the two great groups of his favourite class is divided; so far, then, he is unable to point out any evidence favouring progressive development in it. But, confining himself to species, he sees often before him great varietal changes, so much so as to make it difficult for him to define the species; and it leads him to the belief that such groups were not of independent origin, as was universally thought before Darwin published his great work on the *Origin of Species*. But in this respect the Brachiopoda reveal nothing more than other groups of the organic kingdoms.

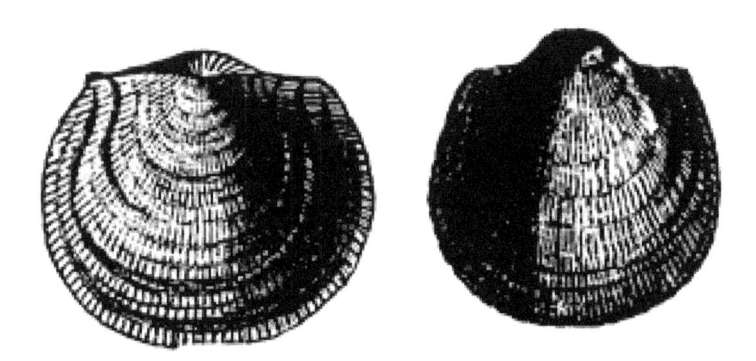

FIG. 60.—*Productus cora* (D'Orbigny). Carboniferous.

"Now, although certain genera, such as *Terebratula, Rhynchonella, Crania*, and *Discina*, have enjoyed a very considerable geological existence, there are genera, such as *Stringocephalus, Uncites, Porambonites, Koninckina*, and several others, which made their appearance very suddenly and without any warning; after a while they disappeared in a similar abrupt manner, having enjoyed a comparatively short existence. They are all possessed of such marked and distinctive internal characters that we cannot trace between them and associated or synchronous genera any evidence of their being either modifications of one or the other, or of being the result of descent with modification. Therefore, although far from denying the possibility or probability of the correctness of the Darwinian theory, I could not conscientiously affirm that the Brachiopoda, as far as I am at present acquainted with them, would be of much service in proving it. The subject is worthy of the continued and serious attention of every well-informed man of science. The sublime Creator of the universe has bestowed on him a thinking mind; therefore all that can be discovered is legitimate. Science has this advantage, that it is continually on the advance, and is ever ready to correct its errors when fresh light or new discoveries make such necessary." The late Joachim Barrande, the great palæontologist of Bohemia, bears similar testimony.

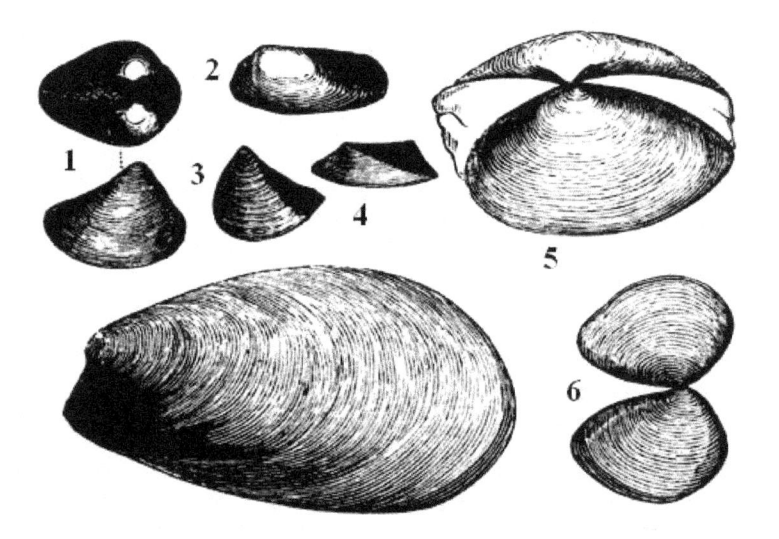

FIG. 61.—Group of Older Palæozoic Lamellibranchs.—After Billings.

1, *Cucullea opima*. 2, *Nucula oblonga*. 3, *Nucula lineata*. 4, *Cypricardia truncata*. 5, *Tellina ovata*. 6, *Nucula bellatula*. 7, *Modiola concentrica*.

The ordinary bivalves, like the mussels and cockles, now so very plentiful on our coasts, are rare in the Cambrian and Silurian, and for the first time make a somewhat conspicuous appearance in the Upper Silurian and Devonian. But from the first they resemble very closely their modern successors, though on the whole neither so large nor so ornate (Fig. 61). Their fortunes have thus been precisely the opposite of those of the Brachiopods, though in neither case is there very marked elevation or deterioration in the individual animals. A very similar statement may be made as to the sea-snails, whether the curious winged snails (Pteropods) or the ordinary crawlers (Gastropods). The former come in early, and are represented by Palæozoic forms finer than any now extant. The genus *Conularia* (Fig. 62) presents some Silurian species six inches or more in length, which are giants in comparison with any now living. The forms of more ordinary Gastropods from the Silurian represented in Fig. 63 will suffice to show that their styles are not very dissimilar from those still extant.[17] As in the case of the ordinary bivalves, however, the modern Gastropods much exceed in numbers and magnitude those of the Palæozoic.

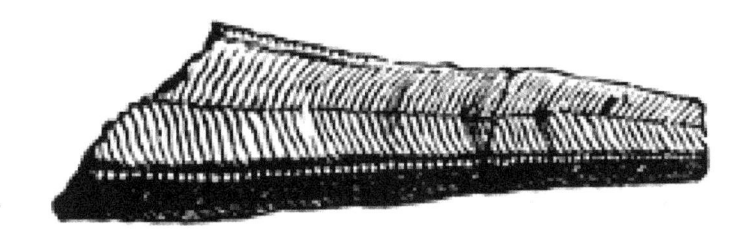

FIG. 62.—*Conularia planicostata* (Dn.). A Carboniferous Pteropod.

FIG. 63.—Silurian Sea-snails. Canada.

a, *Murchisonia bicincta* (Hall). b, *Pleurotomaria umbilicatula* (Hall). c, *Murchisonia gracilis* (Hall). d, *Bellerophon sulcatinus* (Billings).

The highest group of Mollusks, represented in the modern ocean by the Nautili and Cuttle-fishes, has a history so strange and eventful, and so different from what might have been anticipated, that it perhaps deserves a more detailed notice, more especially as Barrande has recently directed marked attention to it in his magnificent work on the Palæontology of Bohemia.

The Cuttle-fishes and Squids and their allies are, in the modern seas, a most important group (Fig. 64). The great numbers in which the smaller species appear on many coasts, and the immense size and formidable character of others; their singular apparatus of arms, bearing suckers, their strange forms, and the inky secretion with which they can darken the water, have at all times attracted popular attention. The great complexity of their structures, and the fact that in many points they stand quite at the head of the invertebrates of the sea, and approach most nearly to the elevation of the true fishes, have secured to them the attention of naturalists. Some of these animals have shelly internal supports, and one genus, that of the Argonauts, or Paper Nautili, has an external protective shell. Allied, though more distantly, to the Cuttle-fishes, are the true Nautili, represented in the modern sea principally by the Pearly Nautilus, though there are two other species, both of them very rare. The modern pearly nautilus (Fig. 65) may be regarded as a peculiar kind of cuttle-fish provided with a discoidal shell for protection, and also for floatage. The shell is divided into a number of chambers by partitions. Of these the animal inhabits the last and largest. The others are empty, and are connected with the body of the animal only by a pipe, or siphuncle, with membranous walls and filled with fluid. Thus provided, the nautilus, when in the water, has practically no weight, and can move up or down in the sea with the greatest facility, using its sucker-bearing arms and horny beak to seize and devour the animals on which it preys. The buoyancy of the shell seems exactly adapted to the weight of the animal; and this proportion is kept up by the addition of new air-chambers as the body increases in size. In the modern seas this singular little group stands entirely isolated, and its individuals are so rare that it is difficult to procure perfect specimens for collections, though its mechanical structure and advantages for the struggle for existence seem of the highest order. But in the old world of past geological time the case was altogether different.

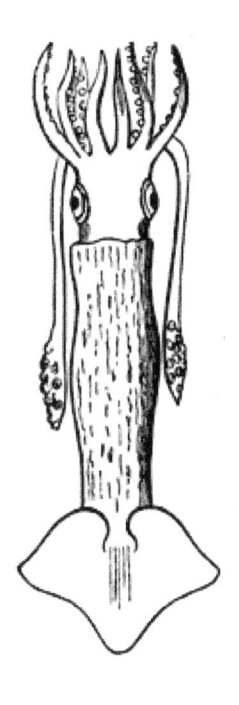

FIG. 65.—Pearly Nautilus (*Nautilus pompilius*).

FIG. 64.—Squid (*Loligo*).

a, Mantle. *b*, Its dorsal fold. *c*, Hood. *o*, Eye. *t*, Tentacles. *f*, Funnel. *g*, Air chambers. *h*, Siphuncle.

The Nautiloid shell-fishes burst suddenly upon us in the beginning of the Siluro-Cambrian, or Lower Silurian, Barrande's second fauna; and this applies to all the countries where they have been studied. In this formation alone about 450 species are known, and in the Silurian these increase to 1,200; and here the group culminates. It returns in the Devonian to about the same number with the Lower Silurian, diminishes in the Carboniferous to 350, and in the Mesozoic, where the Nautiloid forms are replaced by others of the type of the Ammonites, becomes largely reduced. In the Tertiary there are but nineteen species, and, as already stated, in the modern world *three*. These statements do not, however, represent the whole truth. In the Palæozoic, in addition to the genus *Nautilus*, we have a great number of other genera, some with perfectly straight shells, like *Orthoceras* (Fig. 66), others bent (*Cyrtoceras*), others differing in the style of siphuncle, or aperture, or chambers (*Endoceras, Gomphoceras, Lituites*, Figs. 67 to 69), or inflated into sac-like forms (*Ascoceras*). There is, besides, the family of the *Goniatidæ* (Fig. 70), with the chambers thrown into angular folds and the siphuncle at the back. Further, some of the early forms, as the Orthoceratidæ, attain to gigantic dimensions, being six feet or more in length, and nearly a foot in diameter. Thus the idea that we should naturally form from the study of the Nautilus, that it represents a type suited for much more varied and important adaptations than those that we now see, is more than realised in those Palæozoic ages when these animals seem to have been the lords of the seas.

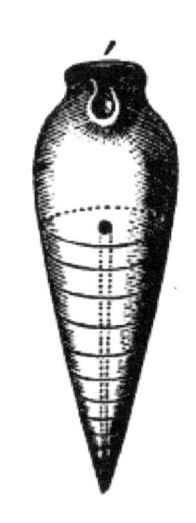

FIG. 67.—*Gomphoceras.*

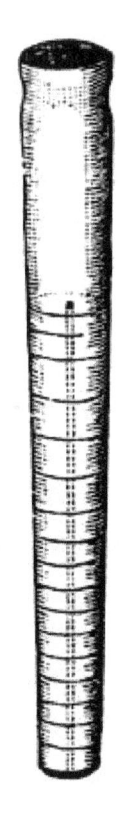

FIG. 66.—*Orthoceras.* Siluro-Cambrian. The dotted line shows the position of the siphuncle.

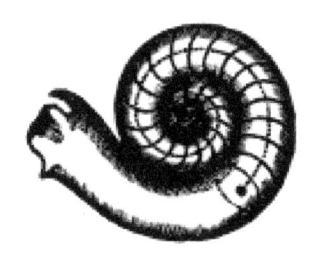

FIG. 68.—*Lituites.*

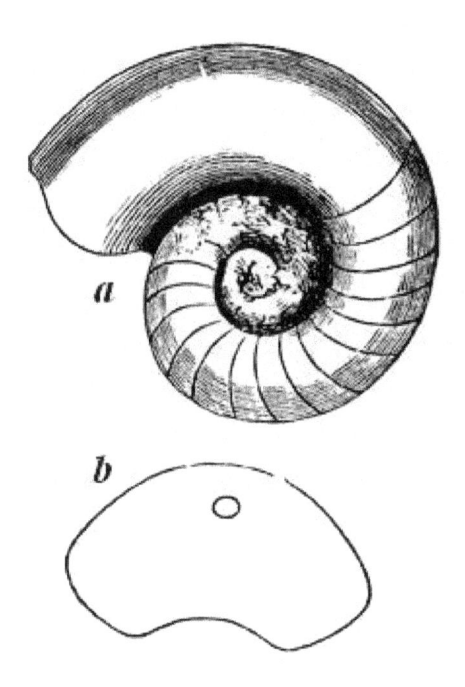

FIG. 69.—*Nautilus Avonensis* (Dn.). Carboniferous.
a, Shell, reduced. *b*, Section, showing siphuncle.

FIG. 70.—*Goniatites crenistria* (Philips). Carboniferous.

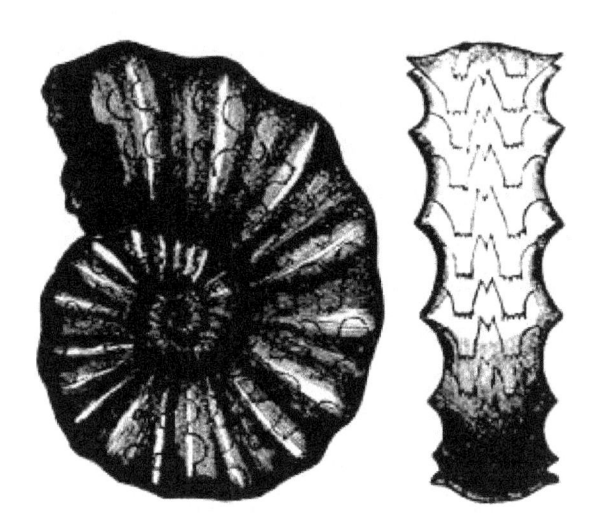

FIG. 71.—*Ceratites nodosus* (Schloth). Triassic.

When we leave the Palæozoic and enter the Mesozoic, though the Nautiloid shells still abound, we find them superseded, in great part, by a nobler form, that of the *Ammonitidæ* (Figs. 71, 72). These are remarkable for the ornate markings on the surfaces of their shells, and for the beautifully waved edges of the partitions (Fig. 72*a*), which, by giving a much more complete support to the sides of the shell, must have contributed greatly to the union of lightness and strength so important to the utility of the shell as a float. This type admits of all the same variety of straight, bent, and curled forms with the simpler Nautiloid type, and some of the species are of great size, Ammonites being known three feet or more in diameter. These animals, unknown in the Palæozoic, appear in numerous species in the Early Mesozoic, culminate in hundreds of beautiful species in the middle of that era, and disappear for ever at its close, leaving no modern successors. Many and beautiful species of Ammonites and their allies have been obtained from the Mesozoic rocks of British Columbia and other parts of the west coast of North America, perfectly representing this group as it occurs at the same period in Europe, and closely resembling the Mesozoic Ammonites of India. These animals have all perished, yet the Atlantic and the Pacific roll between, apparently with conditions as favourable for their comfortable existence as those of any previous time. They perished long ago, at the dawn of the Tertiary; yet the genus Nautilus, one of the oldest and least improved of the whole, survived, and still testifies to the wonderful contrivance embodied in these animals.

FIG. .—*Ammonites Jason* (Reinecke). Jurassic.

FIG. 72a.—Suture of *Ammonites componens* (Meek), of British Columbia. Showing the complicated folding of the edges of the chambers to give strength to the shell. Cretaceous.

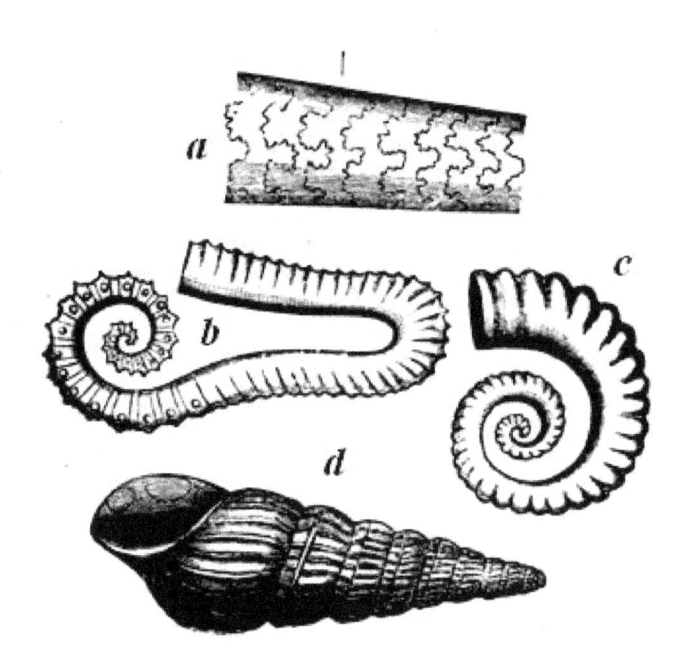

FIG. 73.—Cretaceous Ammonitidæ.

a, Baculites. b, Ancyloceras. c, Crioceras. d, Turrilites.

These are merely general considerations, but Barrande, in his *Études Générales*, goes much farther. He sums up all the known facts in the most elaborate manner, considering first the embryonic characters of the shell in the different genera, then their distribution in space and time, then all the different parts and characters of the shells in the different groups—the whole with reference to any possible derivation of the species; and he finds that all leads to the result that in every respect these shells seem to have been so introduced as to make any theory of evolution with respect to them altogether untenable. In his concluding sentence this greatest of Palæozoic palæontologists affirms that, "The theoretical evolution of the Cephalopods is, like that of the Trilobites, a mere figment of imagination, without any foundation in fact."[18]

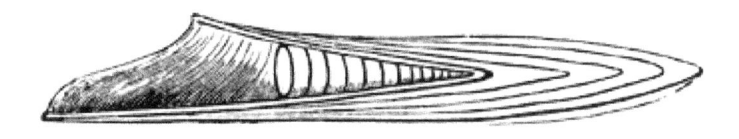

FIG. 74.—*Belemnite.*—After Philips.

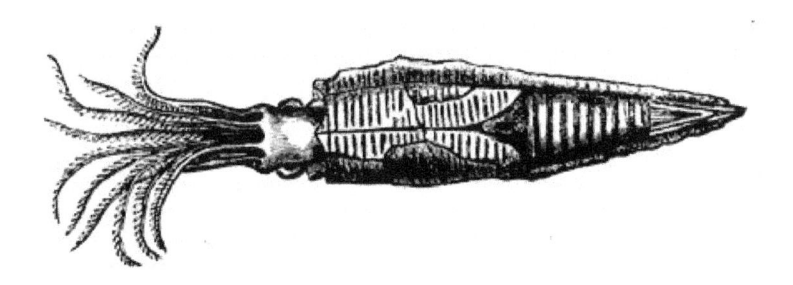

FIG. 74a.—*Belemnoteuthis antiquus.* Supposed to be a Belemnite, with soft parts preserved.—Jurassic.—After Mantell.

I have reserved no space to notice the geological history of the other and higher group of Cephalopods, including the true Cuttles and Squids. This is perhaps less to be regretted, as, from the absence of external shells, they are likely to be much less perfectly known as fossils. So far as known, they are vastly younger than the Nautiloids, for no examples whatever have been found in the Palæozoic. They appear abundantly in the Mesozoic, but are there represented principally by an extinct group of squids (Belemnites and their allies, Figs. 74, 74a), remarkable for the great and complicated development of their internal support, which has a chambered float as well as a solid sheath. This family becomes extinct at the close of the Mesozoic, though the cuttles as a whole perhaps culminate in the modern.

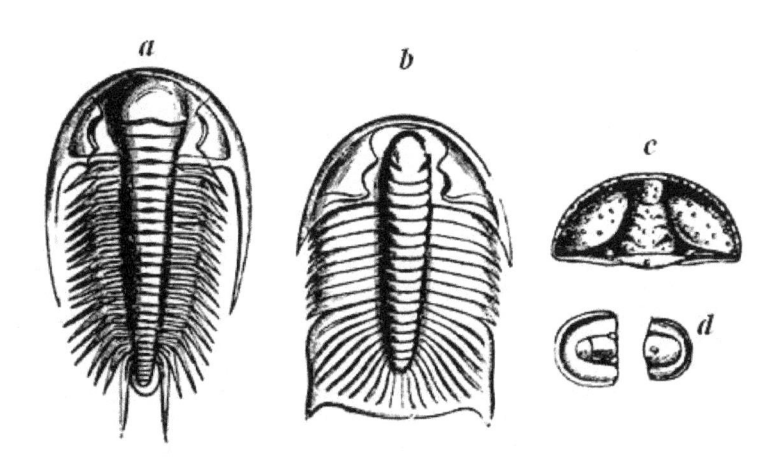

FIG. 75.—Cambrian Trilobites.

a, Paradoxides. *b*, Dikellocephalus. *c*, Conocoryphe (head). *d*, Agnostus (head and tail).

The remarkable group of the Trilobites had precedence in order of time of the Nautiloid shell-fishes. No animal structures can well be more dissimilar than those of the two great groups of aquatic animals which popular speech confounds under the name of "shell-fishes." Take a whelk and a crab, for example, and compare their general forms, the structure of their shells, and their organs of motion, and it is scarcely possible to imagine any two animals more unlike; and when we examine their anatomy in detail this difference does not diminish. They have, it is true, corresponding parts, and these parts serve similar uses, but in plan of structure they are wholly different. Yet both animals may live in the same pool, and may subsist on nearly the same food. If we attempt to find some common type which both resemble, we may trace the structure of the crab back to those of some of the marine worms with which it has some affinity, and those of the whelk to such creatures as the *Lingula*, which are supposed to have a resemblance to the worms. But still the two types, that of the Mollusk and the Articulate, are distinct even from their first appearance in the egg, nor have either any close affinities with the Protozoa, the Hydroids, or the Corals.

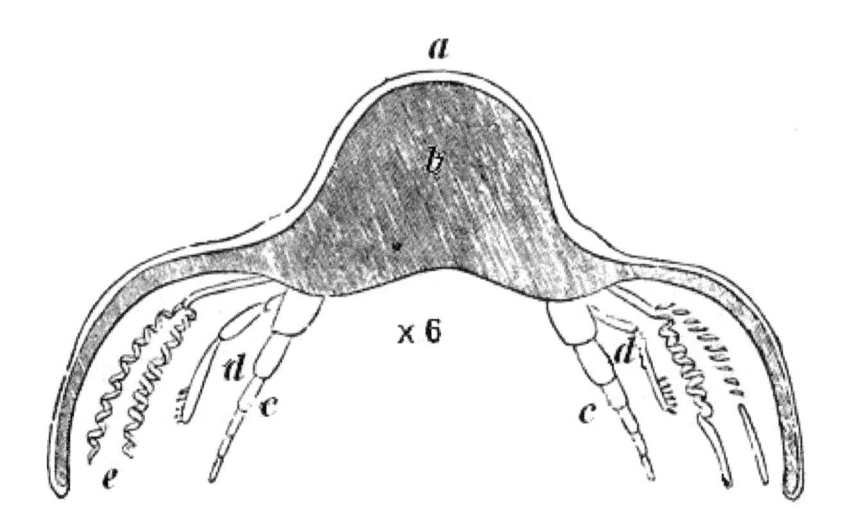

FIG. 76.—Transverse section of *Calymene*. A Silurian Trilobite.—After Wolcott.

a, Dorsal shell. *b*, Visceral cavity. *c*, Legs. *d*, Epipodite—gill-cleaner or palp. *e*, Spiral gills.

Both types meet us in the Early Cambrian, but while the Mollusk is there represented only by low forms, the Articulate is then not only in the humble guise of the worm, but in the complex and highly organised form of the

Trilobite (Figs. 28 and 75). What older phases they may have passed through we know not; but in the Lower Cambrian we have various forms of these animals, including some of the largest known as well as some of the smallest; some of the most complex in number of parts as well as some of the simplest. These animals, in short, seem to have appeared at once all over the world fully formed, and in a variety of generic and specific forms; and nothing short of a very large faith in the imperfection of the geological record can suffice to account for their evolution.

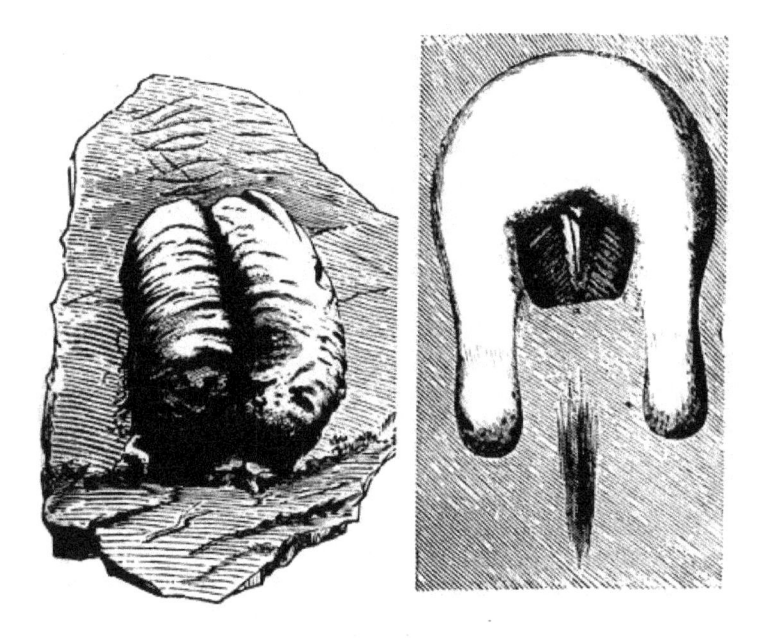

FIG. 76a.—Burrows of Trilobite and of modern King-crab. The Trilobite burrow is known as Ruschinites, and has been supposed to be a sea-weed of the kind called *Bilobites*.

A Trilobite is a creature in whose structure the number three is dominant. Seen from above, it presents three divisions from front to rear:—first, a cephalic shield or head-piece; secondly, a thorax, divided into several segments movable upon each other; and thirdly, a tail-piece or pygidium, which, when brought against the head by the rolling up of the body segments, effectually covers the lower parts. This lower portion was until lately little known; but the discoveries of Billings and of Wolcott have enabled us to restore the jaws under the head, the jointed legs and spiral gills under the thorax, and thus to complete the structure of the animal, and understand better its relations to modern crabs and shrimps (Fig. 76). Of these it certainly comes nearest to the King-crabs and Horseshoe-crabs, a somewhat limited group at present, and one which reaches back in geological time only to the Upper Silurian, when the Trilobites had perhaps already passed their culmination.

Constructed as above described, the Trilobite could swim, as is supposed, usually on its back or side. It could crawl on the bottom. Using its snout as a shovel, it could burrow like a modern King-crab (Fig. 76a); and when pressed by danger some species could roll themselves into balls and defy their enemies.

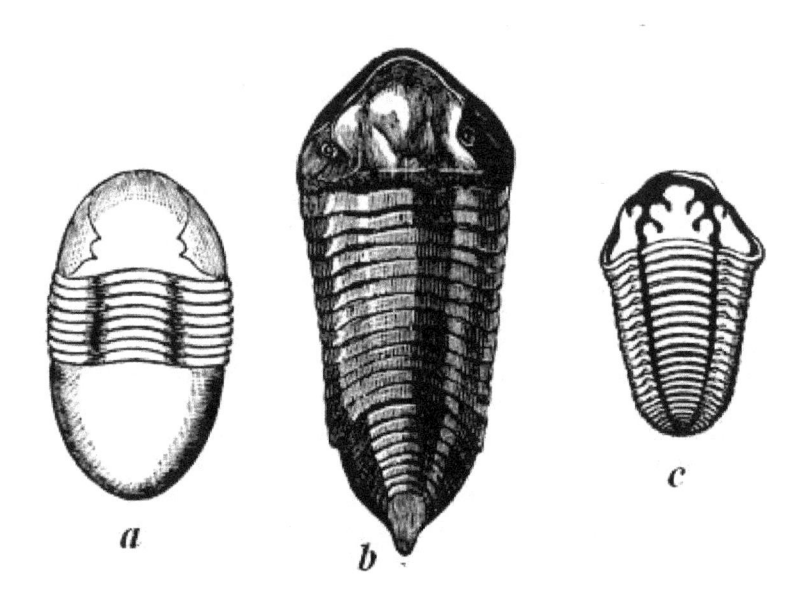

FIG. 77.—Silurian Trilobites.

a, Isotelus. b, Homalonotus. c, Calymene.

This type of animal, entering on the stage in full force in the Older Cambrian, continues under many forms through the whole Palæozoic age, dying out finally in the Carboniferous. Figs. 77 and 78 show a few of the forms of the Silurian, Devonian, and Carboniferous.

Contemporaneously with the dawn of the Trilobite group, appear some small shrimp-like forms (Fig. 28),[19] and others with bivalve shells (Fig. 79), which are closely allied to modern forms,[20] and, like the *Lingulæ*, persist through the succeeding formations with little more than specific change—presenting in this a strange contrast to the Trilobites. While the latter were still flourishing, about the close of the Lower Silurian, a remarkable group of large and highly-developed creatures, allied to the Trilobites, but suited for rapid swimming rather than creeping, was introduced; and in the Upper Silurian and Devonian these creatures[21] attained to gigantic sizes, exceeding, probably, any modern Crustaceans, and were tyrants of the seas. *Pterygotus anglicus* (Fig. 80) is supposed to have attained the length of six feet. Yet these noble representatives of the Crustaceans became extinct in the Carboniferous. On the other hand, a few small king-crabs appear in the Upper Silurian, and this type still continues, and seems to culminate as to size in modern times; so diverse have been the fortunes of these various groups.

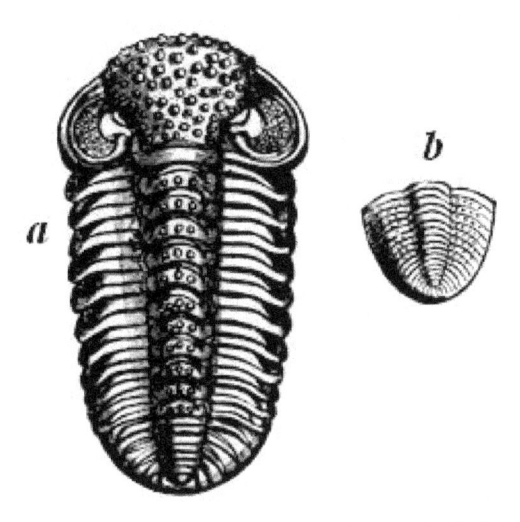

FIG. 78.—Devonian and Carboniferous Trilobites.
a, Phaceps latifrons (Bronn). *b, Philipsia Howi* (Billings) (tail).

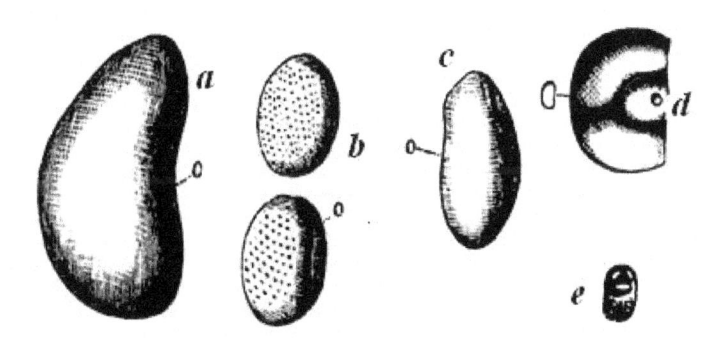

FIG. 79.—Palæozoic Ostracod Crustaceans. Magnified.

a, Bairdia. b, Cytherella inflata (Jones). *c, Cythere.* Carboniferous. *d, Beyrichia Jonesii* (Dn.). Carboniferous. *e, Beyrichia pustulosa* (Hall). Silurian.

The higher, or decapod Crustaceans, now familiar to us in the modern crabs and lobsters, are first found in a few small species in the Devonian[22] and Carboniferous, and they are accompanied in the Devonian by at least one species of the allied group of the Stomapods (Figs. 81, 82).

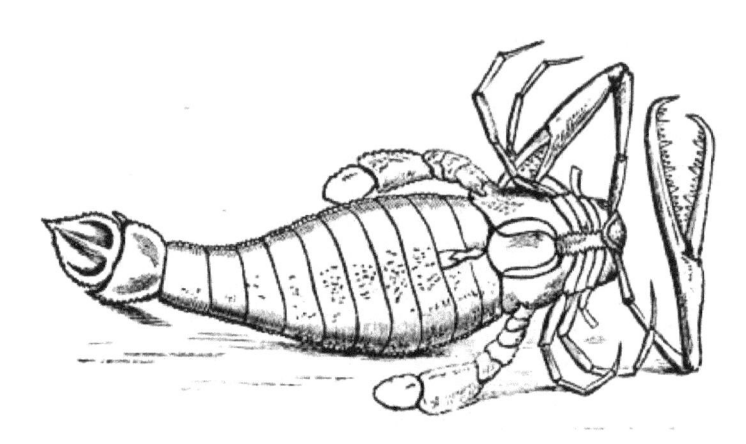

FIG. 80.—*Pterygotus anglicus*. Reduced.—After Page and Woodward.

FIG. 81.—*Amphipeltis paradoxus* (Salter). A Devonian Stomapod from New Brunswick.

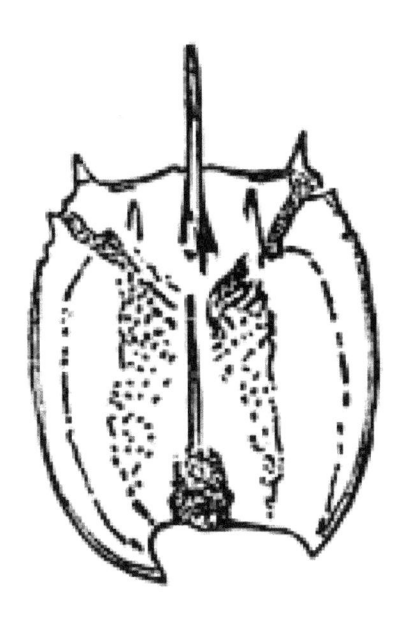

FIG. 82.—*Anthropalæmon Hilliana* (Dn). A Carboniferous Decapod from Nova Scotia. The carapace only.

The Palæozoic age of geology is thus emphatically an age of invertebrates of the sea. In this period they were dominant in the waters, and until toward its close almost without rivals. We shall find, however, that in the Upper Silurian, fishes made their appearance, and in the Carboniferous amphibian reptiles, and that, before the close of the Palæozoic, vertebrate life in these forms had become predominant. We shall also see that just as the leading groups of Mollusks and Crustaceans seem to have had no ancestors, so it is with the groups of Vertebrates which take their places. It is also interesting to observe that already in the Palæozoic all the types of invertebrate marine life were as fully represented as at present, and that this swarming marine life breaks upon us in successive waves as we proceed upward from the Cambrian. Thus the progress of life is not gradual, but intermittent, and consists in the sudden and rapid influx of new forms destined to increase and multiply in the place of those which are becoming effete and ready to vanish away or to sink to a lower place. Farther, since the great waves of aquatic life roll in with each great subsidence of the land, a fact which coincides with their appearance in the limestones of the successive periods, it follows that it is not struggle for existence, but expansion under favourable circumstances and the opening up of new fields of migration that is favourable to the introduction of new species. The testimony of palæontology on this point, which I have elsewhere adduced at length,[23] in my judgment altogether subverts the prevalent theory of "survival of the fittest," and shows that the struggle for existence, so far from being a cause of development and improvement, has led only to decay and extinction, whereas the advent of new and favourable conditions, and the removal of severe competition, are the circumstances favourable to introduction of new and advanced species. This testimony of the invertebrates of the sea we shall find is confirmed by other groups of living beings, to be noticed in the sequel.[24]

NOTE.—The term "Siluro-Cambrian," as used in this and the next chapter, is synonymous with "Ordovician" of Lapworth, which is now coming into somewhat general use.

CORDAITES, OF THE GROUP OF DORY-CORDAITES. BRANCH RESTORED.—After Grand' Eury.

CHAPTER IV.

THE ORIGIN OF PLANT LIFE ON THE LAND.

IF the graphite of the Laurentian rocks was derived from vegetable matter, the further question arises, Was this vegetation of the land, or of the sea? and something may be said on both sides of this question. If there were land plants in the Laurentian period, they must have grown either on rocks older than the Laurentian itself, or on such portions of the beds of the latter as had been raised out of the sea, forming perhaps swampy flats of newly-made soil. But we know no rocks older than the Laurentian, and there is no positive evidence that any of the beds of that formation were other than marine. Still it is not impossible that some of the beds which are now graphitic gneisses may originally have been similar to the bituminous shales, coals, or underclays of the coal formation. The graphite occurring in veins, if of vegetable origin, must have been derived from liquid bitumen oozing into fissures; and veins of this kind occur in later formations, both in marine and fresh-water beds. The only other positive argument which has been adduced in favour of the existence of abundant land plants in the Laurentian is that of Dr. Sterry Hunt, derived from the great beds of iron ore, which it is difficult to account for chemically except on the hypothesis of the decay in the air of great quantities of vegetable matter. The question must remain in doubt till some one is fortunate enough to find portions of the Laurentian carbon retaining traces of organic structure. My own observations, though somewhat numerous, allow me only to say that the graphite sometimes presents fibrous forms, that it occasionally appears as vermicular threads—which, however, I suppose to be fillings of canals of *Eozoon*—and that in the graphitic beds there are occasionally slender root-like bodies of a lighter colour than the mass; but none of these indications are sufficient to determine anything as to its vegetable origin, or the nature of the plants from which it may have been derived.

In any case, the quantity of carbon which has been accumulated in the Laurentian rocks is very great. I have measured one bed at Buckingham, on the Ottawa, estimated to contain 20 per cent. of carbon, and which is at least eight feet in thickness. Sir William Logan has described another similar bed from ten to twelve feet thick, and more recent reports of the Geological Survey of Canada mention a bed supposed to be twenty-five feet thick, in which Mr. Hoffman finds 30 per cent. of carbon. On the whole the quantity of carbon in the graphitic zone of the Laurentian is comparable with that in certain productive coal-fields, and we certainly have in the subsequent geological history no examples of such accumulations except from remains of the luxuriant vegetation of swampy flats.

The Upper Laurentian and Huronian have as yet afforded no evidence of land vegetation. The Cambrian, as already stated, abounds in remains of sea-weeds; but though the forms which have been named *Eophyton* have been regarded as land plants, this claim is, to say the least, very doubtful; and I have as yet seen nothing of this kind which did not appear to me to be merely markings made by objects drifted over the bottom or remains of marine plants. Yet in the Upper Cambrian there are wide surfaces of littoral sandstone often containing minute carbonised fragments, and which might be expected to afford indications of land vegetation, had such existed. I have myself devoted many days of fruitless labour to the examination of the large areas of Potsdam sandstone exposed in some parts of Canada. But as these rocks were evidently formed along the borders of a Laurentian continent capable of supporting vegetation, we may still hope for some discovery of this kind, more especially if we could find the point where some fresh-water stream ran into the Cambrian sea.

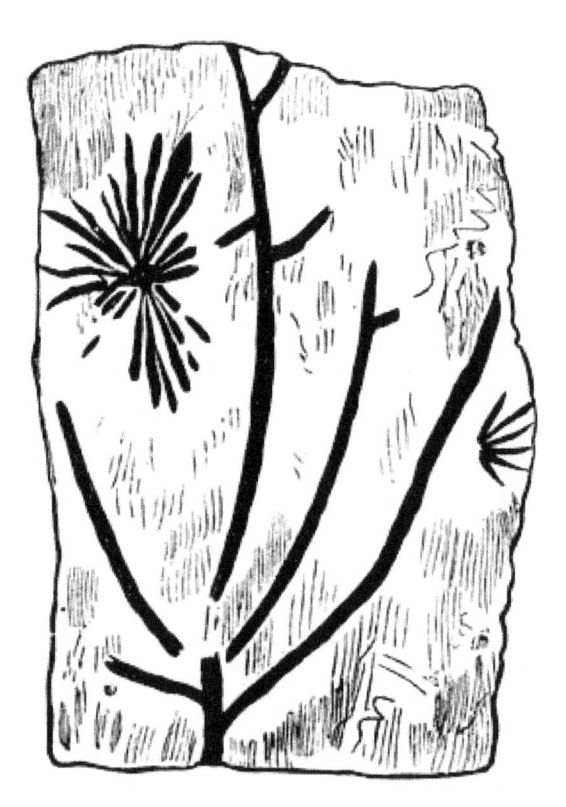

FIG. 83.—*Protannularia Harknessii* (Nicholson). A Siluro-Cambrian Plant, from the Skiddaw series.

The oldest plants, probably higher than Algæ, known to me by their external forms, are those described by Nicholson[25] from the Siluro-Cambrian Skiddaw slates of the north of England (Fig. 83). Their discoverer has named them *Buthotrephis Harknessii* and *B. radiata*,[26] stating, however, that these two species are not improbably portions of the same plant, and that its form is rather that of a land plant than of an Alga. The specimens of these plants which I have seen appear to me to support the conclusion that they represent one species, and this allied to the *Annulariæ* of the Devonian and Carboniferous periods, which probably grew in shallow water with only their upper parts in the air, and bore whorls or verticles of narrow leaves. They were either relatives of the Mare's-tails, or of the Rhosocarps, of our modern swamps and ponds.

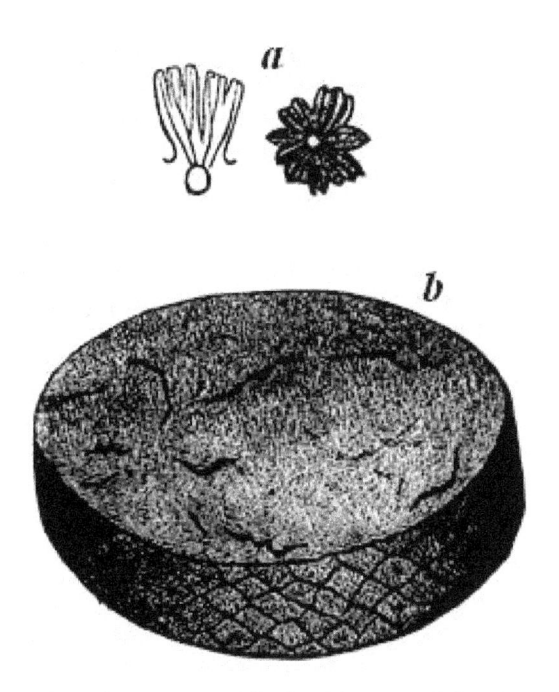

FIG. 84.—American Lower Silurian Plants.—After Lesquereux.

a, *Sphenophyllum primævum*. *b*, *Protostigma sigillarioides*.

Somewhat higher up in the Lower Silurian, in the Cincinnati group of America, Lesquereux finds objects which he refers to the genus *Sphenophyllum*, which is closely allied to *Annularia* (Fig. 84, *a*), and also a plant which he terms *Protostigma* (Fig. 84, *b*), and believes to be the stem of a tree allied to the club-mosses.[27] He also finds minute branching stems, which he refers to the genus *Psilophyton*, to be mentioned in the sequel; but as to these

I have some doubts whether they may not be Zoophytes allied to the Graptolites, rather than plants of that genus. These discoveries tend to show the probable existence in the Siluro-Cambrian of plants representing two of the three leading families of the higher cryptogams or flowerless plants, namely, the Club-mosses and the Mare's-tails. Thus land vegetation begins with the highest members of the lower of the two great series into which botanists divide the vegetable kingdom.

FIG. 86.—Fragment of outer surface of *Glyptodendron* of Claypole. A Silurian Tree.

If we now turn to the Silurian, further evidence of land vegetation presents itself. Near the base of this great series, the club-moss family is represented by a plant discovered by Claypole in the Clinton group, and referred to a new genus (*Glyptodendron*, Fig. 86). Plants of this family have also been noticed by Barrande in Bohemia, and by page in Scotland; and a humble but interesting member of the family, connecting it with the pillworts, *Psilophyton* (Fig. 87), though more characteristic of the Devonian, has been found in the Upper Silurian both in Canada and the United States. No Ferns or Equiseta have as yet been found in the Silurian; but in 1870 I recognised in some fragments of wood from the Ludlow bone-bed, in the Museum of the Geological Survey of Great Britain, the structure of that curious prototypal tree, to which I have given the name *Nematophyton*, and which was first recognised in the Devonian of Gaspé. Since that time I have found in the Upper Silurian beds of Cape Bon Ami, in New Brunswick, similar fragments of fossil wood, associated with round seed-like bodies, having a central nucleus and a thick wall or test of radiating fibres. These bodies show a structure similar to that of those found in the Upper Ludlow of England, and described by Hooker under the name *Pachytheca*. In my judgment they are certainly true seeds.[28] Seeds of this kind have also been found by Hicks in the still older

Denbighshire grits of North Wales, along with fragments of the wood of *Nematophyton*, and with remains of branching stems which have been described under the name *Berwynia*, though it is not unlikely that they represent the branches of *Nematophyton*. It is proper to add that these ancient vegetable fossils are regarded by some English botanists as gigantic algæ or sea-weeds, but I confess I am unable to adopt this view of their nature. The supposed fern of the Upper Silurian, figured in the first edition of this work, has proved on further examination to be merely an imitative form produced by crystallisation. On the other hand, the recent discovery of a cockroach and two species of Scorpion in the Silurian, proves the existence of land animals as well as plants at this period.

FIG. 87.—*Psilophyton princeps* (Dn.) Silurian and Devonian. Restored.

a, Fruit, natural size. *b*, Stem, natural size. *c*, Scalariform tissue of the axis, highly magnified. In the restoration one side is represented in vernation, and the other in fruit.

It is probable that these discoveries represent merely a small proportion of the plants actually existing in the Silurian period. All the deposits of this age at present known to us are marine; and most of them were probably formed at a distance from land, so that it is little likely that land plants could find their way into them. At any time the discovery of an estuarine or lacustrine deposit of Silurian age might wonderfully extend our knowledge of this ancient flora.

The Devonian or Erian age, that of the classic Old Red Sandstone of Scotland, is that in which we find the first great and complete land flora; and though this is inferior in number of species to that of the succeeding Carboniferous, and greatly less important with reference to its practical bearing on our welfare, it is in some respects superior in that variety which depends on diversity of soil and of station. To appreciate this, it will be necessary to glance at the range and subdivisions of the modern flora.

In the modern world we divide all vegetation into two great series, that of the Flowering Plants (*Phænogams*), which also produce true fruits and seeds, and that of the Flowerless Plants (*Cryptogams*), which produce minute spores instead of seeds. The latter is in every respect the lower group. This lower series is again divisible into three classes—first and lowest, that of the Seaweeds, Moulds, and Lichens (Thallophytes). Secondly, that of the Mosses and their allies (Anophytes). Thirdly, that of the Ferns, Equisetums and Club-mosses (Acrogens). In like manner the second, or higher series is divisible into three classes: that of the Pines and Cycads (Gymnosperms), having naked seeds not covered by true fruits, and woody tissue of simple structure; that of the Palms and Grasses and their allies (Endogens); and last and highest, that of the ordinary timber trees and other plants allied to them, with exogenous stems, netted-veined leaves, and a two-leaved embryo (Exogens). These last are in every respect the dominant plants on our present continents. Carrying with us this twofold division of the vegetable kingdom and its subdivisions, we shall be prepared to understand the relation of the more ancient floras to that now living.

FIG. 88.—Trunk of a Devonian Tree-fern (*Caulopteris Lockwoodi*, Dn.). Gilboa, New York. One-third natural size.

FIG. 89.—Frond of *Archæopteris Jacksoni* (Dn.). Devonian, of Maine.

- 90 -

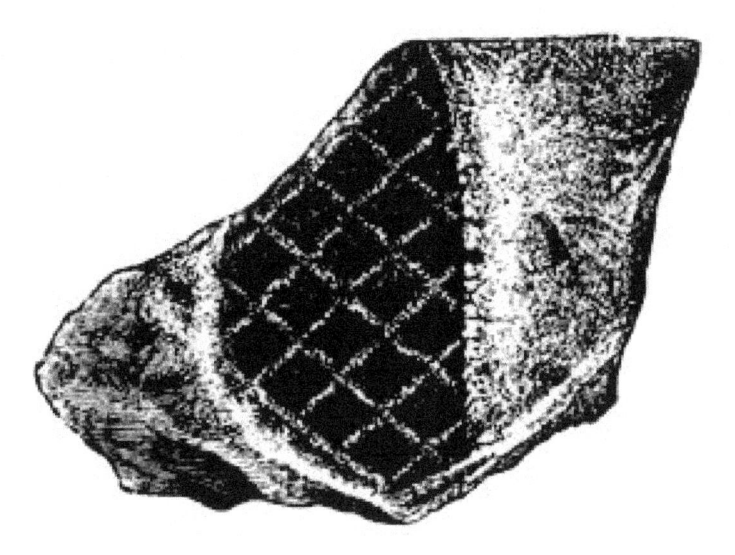

FIG. 90.—Portion of a branch of *Leptophleum rhombicum* (Dn.). A Lycopodiaceous tree of the Devonian of Maine. Natural size.

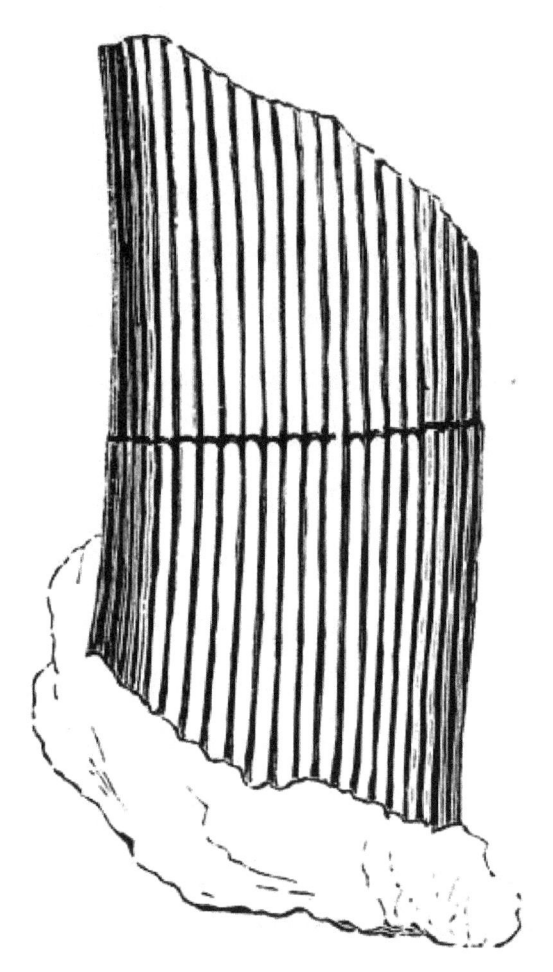

FIG. 91.—*Calamites radiatus* (Brongniart). Middle Devonian of N. Brunswick.

In the Devonian age we meet with no land plants of the two lower classes of the Cryptogams, and with scarcely any that can be referred to the two higher classes of Phænogams, so that the vegetation of this period presents a remarkable character of mediocrity, being composed almost entirely of the highest class of the flowerless plants and the lowest class of those that flower. Of the former there are Tree-ferns and vast numbers of herbaceous forms (Figs. 88, 89), great Lycopodiaceous plants, immensely better developed than those now existing (Fig. 90), and gigantic *Calamites*, allied to the Mares'-tails (Fig. 91), along with humbler members of the same group (Fig. 95). Of the latter there were Pines of great stature, known to us at present only by their wood (Fig. 92); and that other allied trees existed we have evidence in numerous seeds which must have belonged to this class (Fig. 93), and in long flag-like leaves[29] which modern discoveries would refer to the same group. As yet we know no Devonian Palms or Grasses; and only a single specimen has been found indicating the existence of a plant of the highest vegetable class, that of the true exogens. This unique specimen, found by Hall in the Devonian of the shores of Lake Erie, is a fragment of mineralised wood, the structures of which are represented in Fig. 94. The large ducts seen in cross section in Nos. 1, 2, and 3, and in longitudinal section in Nos. 4 and 5, and the medullary rays, seen in Nos. 1, 4, and 6, testify to the fact that this chip of wood must have belonged to a tree of the same type which contains our oaks, maples, and poplars; a type which does not appear to have become dominant till near the close of the Mesozoic, but which already existed, though perhaps only in few species, and only in upland and inland positions, as far back as the Middle Devonian. Perhaps one of the most interesting discoveries in the Erian or Devonian rocks has been that of the immense abundance of spores of those humble plants the Rhizocarps, represented in modern times by the Pillworts and Salviniæ, &c. To these it is believed that *Sphenophyllum* and *Psilophyton* were allied; but in addition to this there are thick and vastly extended beds of bituminous shale which owe their inflammable properties to countless multitudes of Macrospores (*Sporangites*) of the genus *Protosalvinia*.[30] In Ohio there are beds of this kind 350 feet thick, and extending across the State. They occur also in Canada, where these forms were first recognised by the writer in the bituminous shale of Kettle Point, Lake Huron.

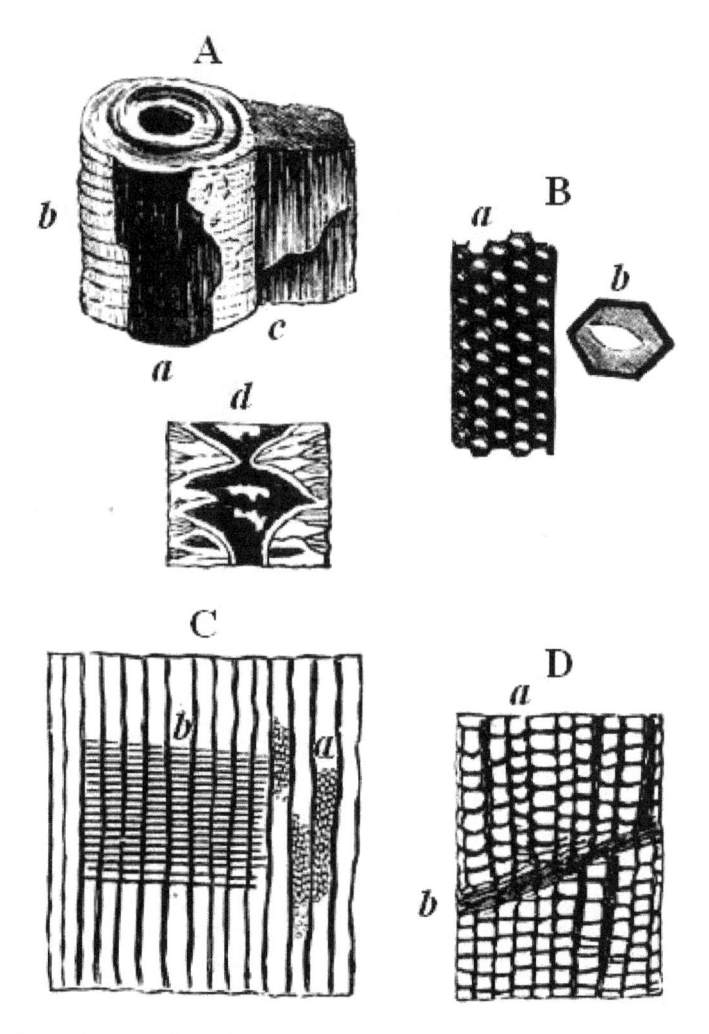

FIG. 92.—A Devonian Taxine Conifer (*Dadoxylon ouangondianum*, Dn.). St. John, New Brunswick.

A, Fragment showing *Sternbergia* pith and wood; *a*, Medullary sheath; *b*, Pith; *c*, Wood; *d*, Section of pith.

B, Wood cell *a*, and hexagonal areole and pore *b*.

C, Longitudinal section of wood, showing *a*, Areolation, and *b*, Medullary rays.

D, Transverse section showing *a*, Wood-cells, and *b*, Limit of layer of growth.

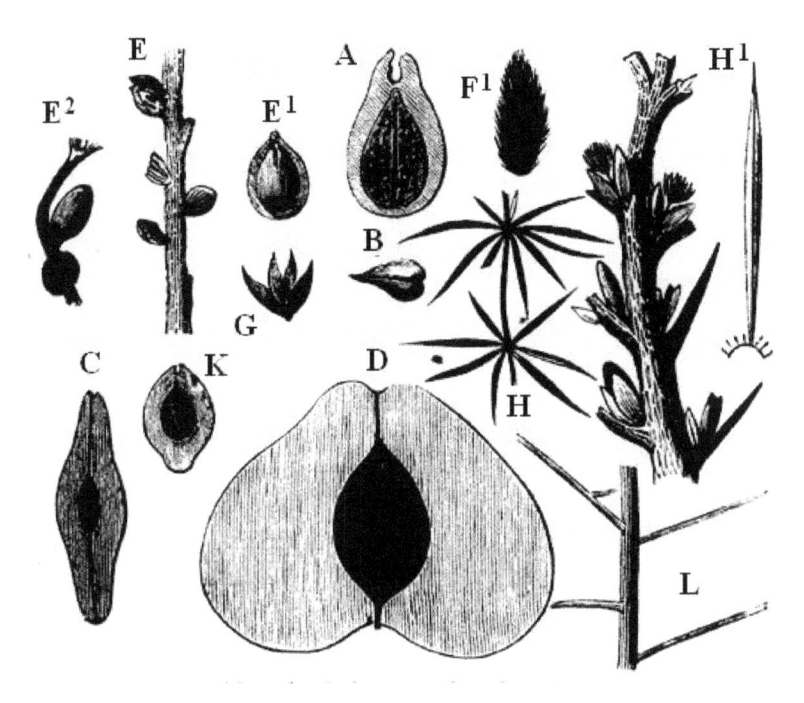

FIG. 93.—Group of Devonian Fruits, &c. Middle Devonian, New Brunswick.

A, *Cardiocarpum cornutum.*

B, *Cardiocarpum acutum.*

C, *Cardiocarpum Crampii.*

D, *Cardiocarpum Baileyi.*

E, *Trigonocarpum racemosum.*

E¹, E², Fruits enlarged.

F, *Antholithes Devonicus.*

F¹, Fruit of the same.

G, *Annularia acuminata.*

H, *Asterophyllites acicularis*　　H¹, Leaf.

K, *Cardiocarpum.* (? young of A.)

L, *Pinnularia dispalans.*

FROM *Acadian Geology.*

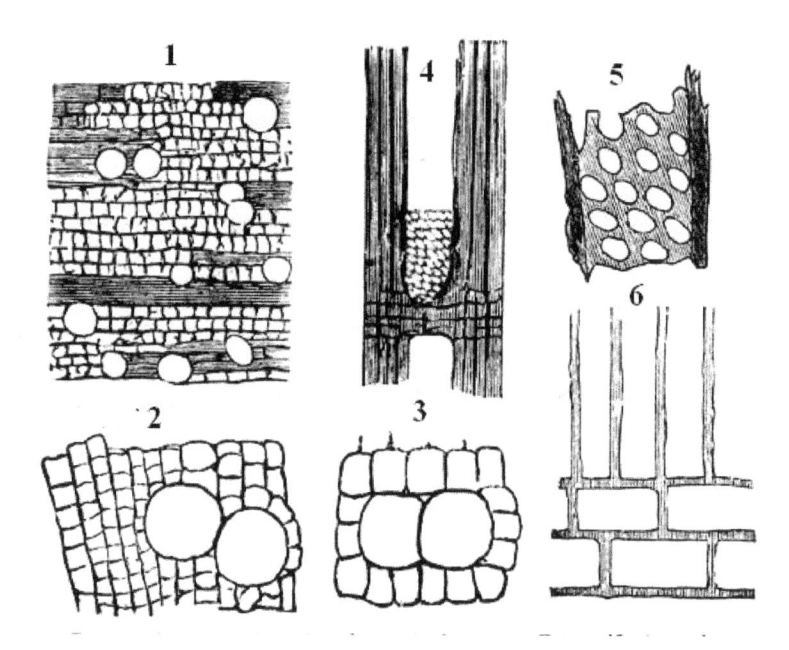

FIG. 94.—Structures of the oldest-known Angiospermous Exogen (*Syringoxylon mirabile*, Dn.). From Eighteen-mile Creek, Lake Erie.

1, Transverse section x 100. 2 and 3, Portions of the same x 300. 4, Longitudinal section x 300. 5, Fragment of duct from the same x 600. 6, Wood cells and medullary ray x 600.

The Devonian flora seems to have been introduced in the northern parts of the American continent at a time of warm and equable climate, and of elevation of new land out of the Silurian sea. It spread itself to the southward, and was finally destroyed in the great subsidences and disturbances which closed the Devonian age, and which were probably accompanied with refrigeration of climate. It was succeeded by the more massive and richer, but more monotonous flora of the Carboniferous, a period in which large areas of our continents were in the state of swampy and often submerged flats, and in which the climate was again warm and uniform.

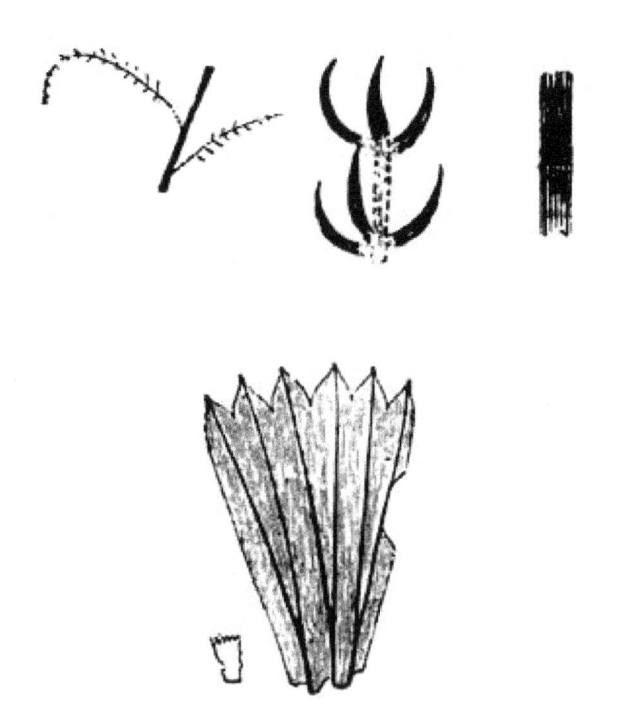

FIG. 95.—*Asterophyllites parvula* (Dn.), and *Sphenophyllum antiquum* (Dn.). Middle Devonian, New Brunswick.

The Carboniferous age was, even more emphatically than the Devonian, an age of Acrogens and Conifers. A few Carboniferous Fungi have recently been discovered, but there are no known Lichens or Mosses. There seem to be a few Endogens, but no true Exogens. The great bulk of the plants consists of Acrogens and Gymnosperms, as in the previous period. As this flora is so very important and so much better known than any other of those belonging to the infancy of the vegetable kingdom, we may notice a little in detail some of its leading forms.

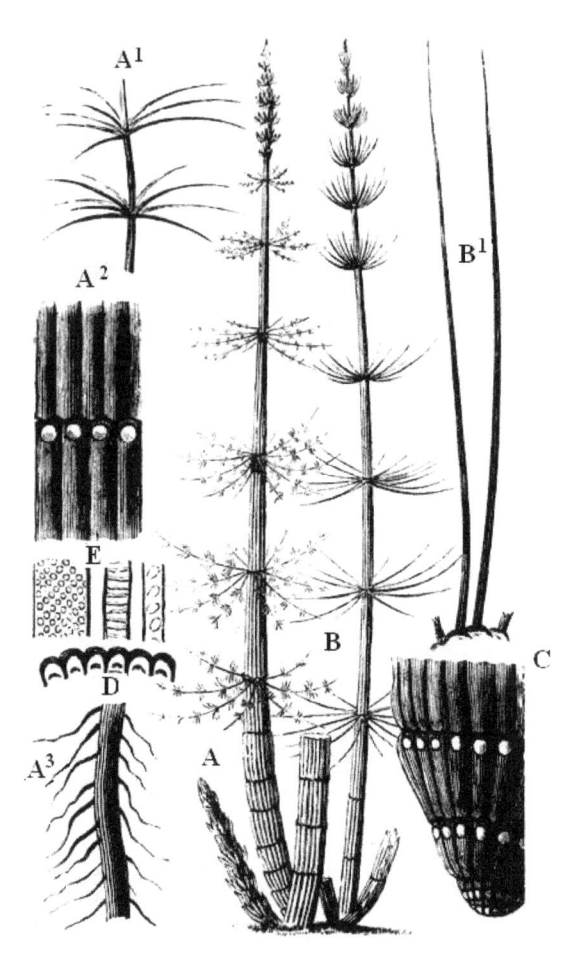

FIG. 96.—*Calamites*. Carboniferous.

A, *C. Suckovii*. B, *C. Cistii* (Bt.). C, Base of *Calamites*. D, E, Structures. FROM *Acadian Geology*.

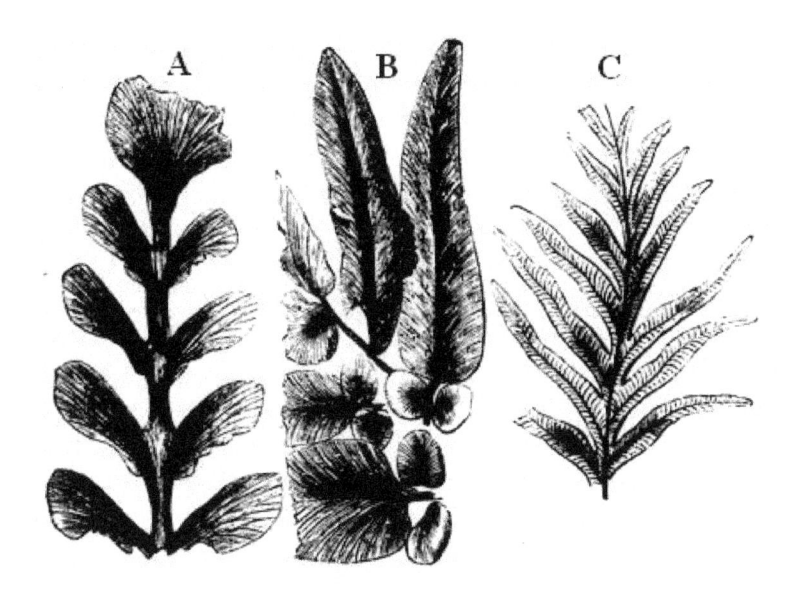

FIG. 97.—Carboniferous Ferns.

A, *Odontopteris subcuneata* (Bunbury). B, *Neuropteris cordata* (Brongniart). C, *Alethopteris tonchitica* (Brongniart).

Beginning with the Mares'-tails, we find these represented in the Carboniferous by many gigantic species, attaining to almost tree-like dimensions (Fig. 96). These are the *Calamites*, which formed dense brakes and jungles on the margins of the great swampy flats of this period. Their tall stems, ribbed and jointed, bore whorls of leaves or branchlets. Sending out horizontal root-stocks and budding out from the base, they grew in great clumps, and had the capacity to resist the effects of accumulating sediment by constantly sending out new stems at higher and higher levels. The larger species assumed a complexity in the structure of their stems unknown in their modern congeners, and enabling them to grow to a great height;[31] but their foliage and fructification were not correspondingly advanced. Thus the family of the Equisetaceæ culminated in the Carboniferous, and thenceforth descended gradually in the succeeding ages, leaving the comparatively humble Mares'-tails and Scouring Rushes as its present representatives.

The Ferns of the Carboniferous, like those of the Devonian, presented both gigantic forms like those of the tree-ferns of the modern tropics, and delicate herbaceous species, and these in great profusion. On the whole, they do not strike the observer as very dissimilar from those of modern times. A more critical examination, however, shows that the bulk of the tree-ferns of the Devonian and Carboniferous are allied not to the Polypod type, which is the most common at present, but to certain comparatively rare southern ferns, the *Marattias* and their allies, characterised by a peculiar style of fructification, perhaps adapting them to a moist and warm atmosphere (Fig. 97).[32] Thus the ferns, while a wonderfully persistent type, were in their grander forms far more widely distributed in the Carboniferous than at present; and genera now comparatively rare, and limited to warm and moist climates, were then abundant, and ranged over those temperate and boreal regions of the Northern Hemisphere where only a few humble and hardy species can now subsist. There were also some remarkable and anomalous tree-ferns, of which that represented in Fig. 98 is an example.

FIG. 98.—Carboniferous Tree-ferns.

A, *Megaphyton magnificum* (Dn.). C, *Palæopteris Hartii* (Dn.). D, *P. Acadica* (Dn.).

The family of the Club-mosses, already, even in the Devonian, in advance of its modern development, experiences in the Carboniferous a remarkable and portentous extension into great trees of several genera and many species, constituting apparently extensive forests, and having the woody tissues of their stems developed to a degree unheard of in their present representatives (Fig. 99). Further, they become closely linked, in external form at least, with another and more advanced type, that of the *Sigillariæ*. These remarkable trees were the most abundant of all in the swamps of the coal-formation, and probably those which most contributed to the accumulation of coal. They presented tall pillar-like trunks, often ribbed longitudinally, and with perpendicular rows of scars of fallen leaves. Dividing at top into a few thick branches, they were covered with long rigid grass-like foliage. Their fruit was borne in rings or whorls of spikes surrounding the branches at intervals (Fig. 100). Their roots were strangely symmetrical, spreading out like underground branches into the soft soil by a regular process of bifurcation, and were covered with rootlets diverging in every direction, and so jointed to the main root that when broken off they left round marks regularly arranged. These

roots are the so-called *Stigmariæ*, so abundant in every coal-field, and especially filling the "under-clays" of the coal-beds, which are the soils on which the plants forming these beds were supported. The true botanical position of the *Sigillariæ* has been a matter of much controversy. Some of them undoubtedly have structures akin to those of the tree-like Club-mosses, as Williamson has well shown, and may have been cryptogamous. Others have structures of higher character, akin to those of the modern Cycads, and seem to have borne nutlets allied to those of these plants. Yet the external forms of these diverse sorts are so similar that no definite separation of them has yet been made. Either these anomalous trees constitute a link connecting the two great series of the vegetable kingdom, or we have been confounding two distinct groups, owing to imperfect information.

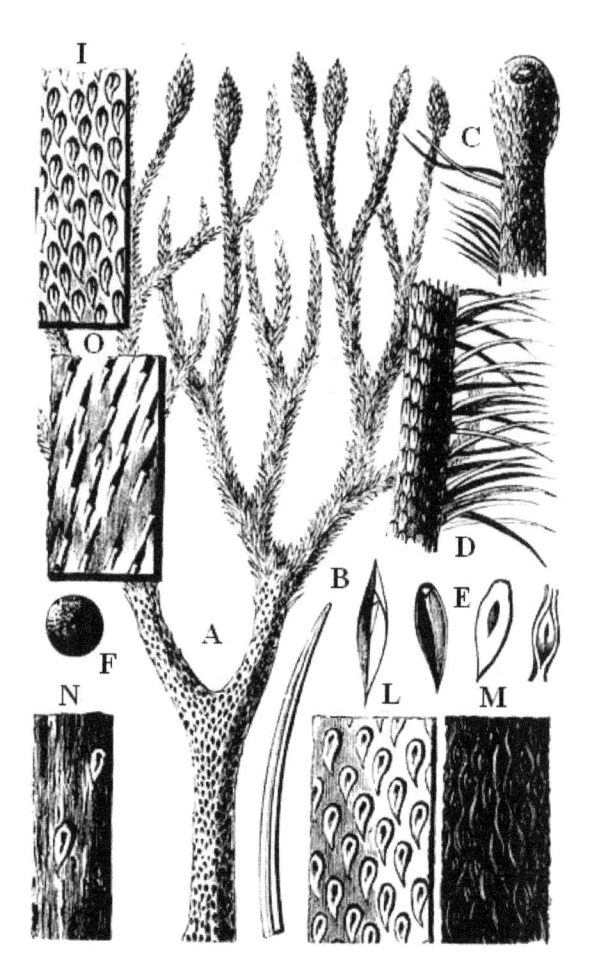

FIG. 99.—*Lepidodendron corrugatum* (Dn.). A characteristic Lycopod of the Lower Carboniferous of America.

A, Restoration. B, Leaf, natural size. C, Cone. D, Leafy branch. E, Forms of leaf-bases. F, Sporangium. I, L, M, N, O, Markings on stem and branches, in various states.

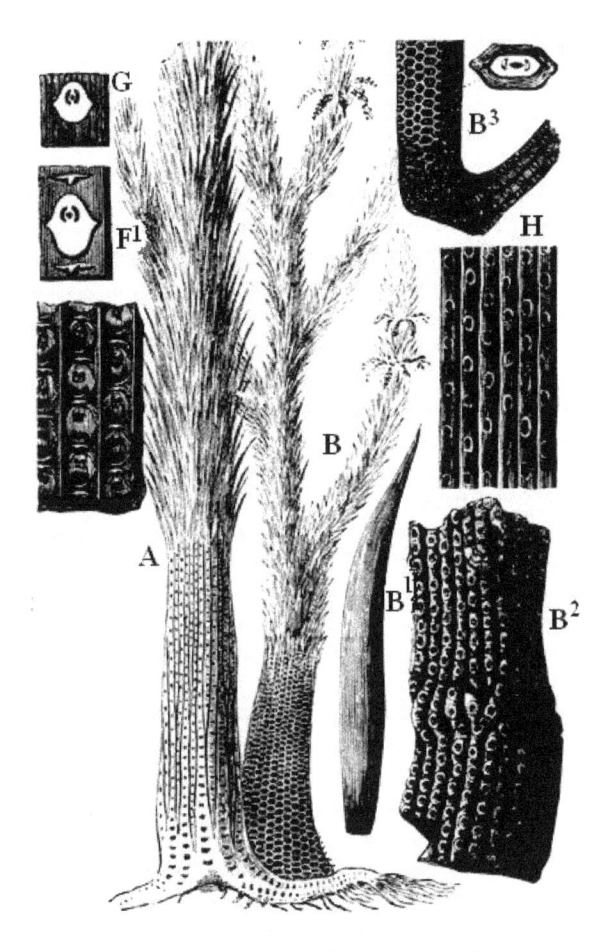

FIG. 100.—*Sigillariæ* of the Carboniferous.

A, *Sigillaria Brownii* (Dn.). B, *S. elegans* (Brongniart). B¹, &c. Leaf and Leaf-scars.

Another curious, and till recently little understood, group of Carboniferous trees is that known as *Cordaites*, which existed already in some of its species in the Devonian. Their leaves are long, and often broad as well, and with numerous delicate parallel veins, resembling in this the leaves of grasses. Corda long ago showed that one species at least has a stem allied to the Clubmosses. More recently Grand' Eury has found in the South of France admirably preserved specimens, which show that others more resembled in their structure the Pines and Yews, and were probably Gymnosperms, approaching to the Pines, but with very peculiar and exceptional foliage, of which the only modern examples are the broad-leaved Pines of the genus *Dammara* (Frontispiece to Chapter). Here again we have either two very distinct groups, combined through our ignorance, or a connecting link between the Lycopods and the Pines.

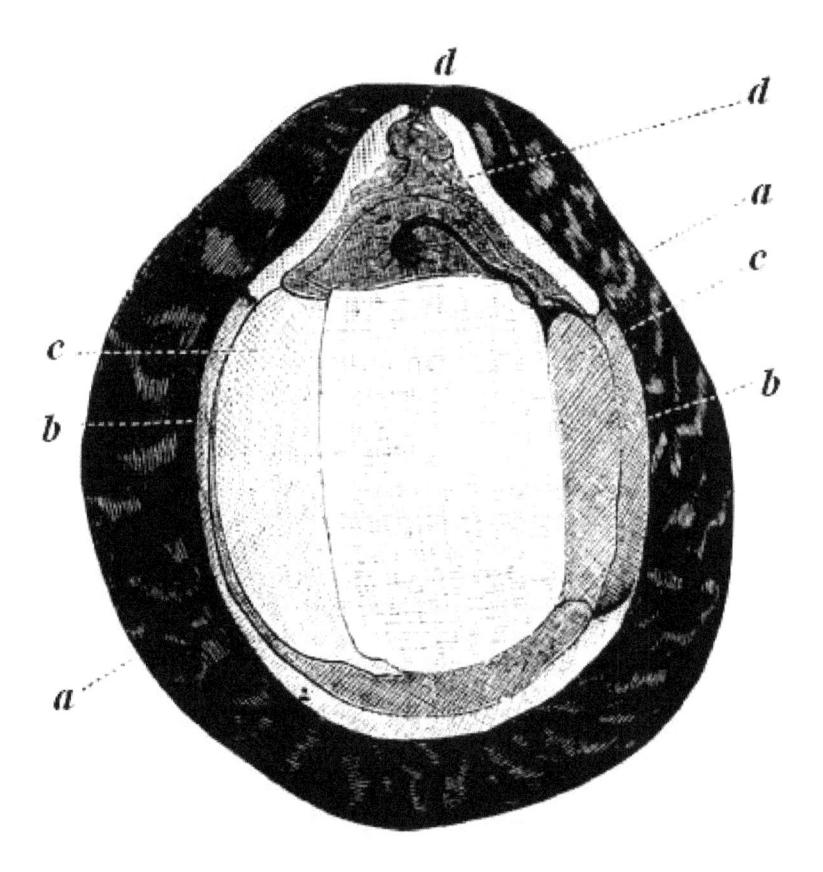

FIG. 101.—*Trigonocarpum Hookeri* (Dn.). A Gymnospermous seed.

a, Testa. *b*, Tegmen. *c*, Nucleus. *d*, Embryo.

The Yews and their allies among modern trees, while members of the great Cone-bearing order, bear nut-like seeds in fleshy envelopes, sometimes, as in the Ginkgo of Japan, constituting edible fruits. Seeds of this type seem to have been extremely abundant in the Carboniferous age in all parts of the world, and were probably produced by trees of several genera (*Dadoxylon, Sigillaria, Cordaites,* etc.) (Fig. 101). Charles Brongniart has recently described no less than seventeen genera of these seeds from the coal-field of St. Étienne alone, and it would be a low estimate to say that we probably know as many as sixty or seventy species in all, while the trunks of great coniferous trees allied to Taxineæ, and showing well-preserved structure, are by no means uncommon in the Devonian and Carboniferous. Had these great

Yews appeared for the first time in the Coal-formation, we might have supposed that they had been developed from such Lycopods as Lepidodendra, and that the *Cordaites* are the intermediate forms; but unfortunately the Pines go almost as far back in geological time as the Lycopods, and it does not help us, when in search of evidence of evolution, to find the link which is missing or imperfect in the Early Devonian supplied in the Coal-formation, where, for this purpose at least, it is no longer needed.

We have said something of what was in the Palæozoic flora; but what of that which was not? We may answer:—Nearly all that is characteristic of our modern forests, whether in the ordinary Exogens, which predominate so greatly in the trees and shrubs of temperate climates, or in the Palms and their allies, which figure so conspicuously within the tropics. The few rare, and to some extent doubtful, representatives of these types scarcely deserve to be noted as exceptions. Had a botanist searched the Palæozoic forests for precursors of the future, he would probably have found only a few rare species, while he would have seen all around him the giant forms and peculiar and monotonous foliage of tribes now degraded in magnitude and structure, and of small account in the system of nature.

It must not be supposed that the Palæozoic flora remained in undisturbed possession of the continents during the whole of that long period. In the successive subsidences of the continental plateaux, in which the marine limestones were deposited, it was to a great extent swept away, or was restricted to limited insular areas, and these more especially in the far north, so that on re-elevation of the land it was always peopled with northern plants. Thus there were alternate restrictions and expansions of vegetation, and the latter were always signalised by the introduction of new species, for here, as elsewhere, it was not struggle, but opportunity, that favoured improvement.

In the Lower Silurian such plants as existed must have experienced great restriction at the age of the Niagara or Wenlock limestone. Those of the Upper Silurian suffered a similar reverse at the time of the Lower Helderberg or Ludlow limestones. This recurred at the close of the Devonian and in the time of the Lower Carboniferous limestone; and finally the Palæozoic flora disappeared altogether in the Permian, to be replaced by new types in the Mesozoic. While, therefore, there is a great general similarity in the successive Palæozoic floras, there are minor differences, so that the Devonian plants are for the most part distinct specifically from those of the Lower Carboniferous, those of the Lower Carboniferous from those of the Coal-formation, and those of the latter from those of the Permian.

With all these vicissitudes it is to be observed that there is no apparent elevation of type in all the long ages from the Devonian to the Permian, that the Acrogens and Gymnosperms of these periods are in some respects superior, in all respects equal, to their modern successors, and that their history shows a decadence toward the modern period; that intermediate forms arrive too late to form connecting links in time, that several distinct types appear together at the beginning, and that all utterly and apparently simultaneously perish at the end of the Palæozoic, to make way for the entirely new vegetation of the succeeding age. Theories of evolution receive no support from facts like these, though their practical significance, as parts of the one great uniform scheme of nature, is sufficiently manifest.

Of what use then were these old floras? To the naturalist, vegetable life, with regard to its modern uses, is the great accumulator of pabulum for the sustenance of the higher forms of vital energy manifested in the animal. In the Palæozoic this consideration sinks in importance. In the Coal period we know few land animals, and these not vegetable feeders, with the exception of some insects, millipedes, and snails. But the Carboniferous forests did not live in vain, if their only use was to store up the light and heat of those old summers in the form of coal, and to remove the excess of carbonic acid from the atmosphere. In the Devonian period even these utilities fail, for coal does not seem to have been accumulated to any great extent, and the petroleum of the Devonian appears to have been produced from aquatic vegetation. Even with reference to theories of evolution, there seems no necessity for the long continuance and frequent changes of species of acrogenous plants without any perceptible elevation. We may have much yet to learn of the life of the Devonian; but for the present the great plan of vegetable nature goes beyond our measures of utility; and there remains only what is perhaps the most wonderful and suggestive correlation of all, namely, that our minds, made in the image of the Creator, are able to trace in these perished organisms structures similar to those of modern plants, and thus to reproduce in imagination the forms and habits of growth of living things which so long preceded us on the earth. We may indeed proceed a step further, and hold that, independently of human appreciation, these primitive plants commended themselves to the approval of their Maker, and perhaps of higher intelligences unknown to us; and that in the last resort it was for His pleasure that they were created.

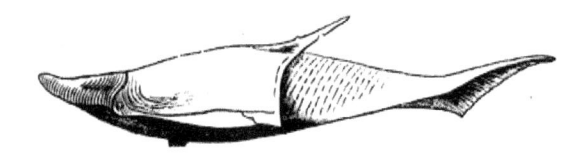

PTERASPIS. RESTORED.—AFTER LANKESTER.

CHAPTER V.

THE APPEARANCE OF VERTEBRATE ANIMALS.

CONFESSEDLY the highest style of animal is that which possesses a skull and backbone, with brain and nerve system to match, and which embodies the general plan of structure employed in man himself. Yet among the fishes, which constitute the lowest manifestation of this type, are some so rudimentary that the brain is scarcely developed, and the skeleton is merely a cord of gristle. These are represented in the modern world only by the Lancelot,[33] a creature which has sometimes been mistaken for a worm, and by a slightly more advanced type, that of the Lampreys.[34] In these animals the Vertebrates make the nearest approach to the lower domains of the animal kingdom, collectively known as Invertebrates. We should naturally expect that since the vertebrates succeed the inferior animals in time, their lower types should appear first, and that these should be aquatic rather than terrestrial. On the other hand, as the oldest fishes that are certainly known are strongly protected with bony armour, and had to contend against formidable Crustaceans and Cuttles, we might suppose that the Lancelot and the Lampreys are rather degraded types belonging to the modern period, than the true precursors of the other fishes.

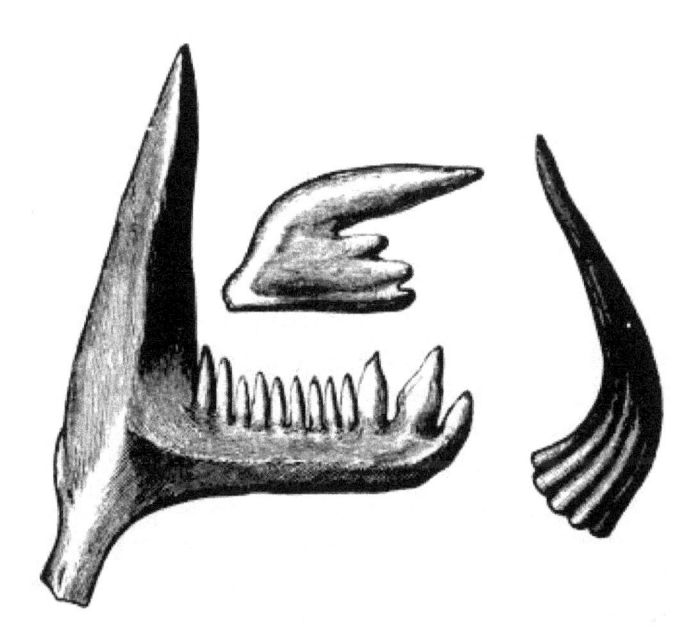

FIG. 102.—Siluro-Cambrian Conodonts. Magnified.—After Pander.

But if fishes like the Lancelot preceded all others, we may never find in a fossil state any traces of their soft and perishable bodies; and even the Lampreys have no hard parts except small horny teeth, which might easily escape observation. But palæontologists have sharp eyes, and it has not escaped them that certain microscopic tooth-like bodies are somewhat widely distributed in the older rocks. In Russia, Pander has found in the Upper Cambrian and Lower Silurian, and also in the Devonian and Carboniferous, minute conical and comb-like teeth, to which he has given the name of *Conodonts* (Fig. 102), and which he supposes to be the teeth of ancient Lampreys. Similar teeth have been found by Moore and others in the Carboniferous of England, and by Newberry in Carboniferous shales in Ohio. In point of form, these bodies certainly resemble the teeth of the humble fishes to which they have been referred. In the case of the Carboniferous specimens from Ohio—the only ones I have had an opportunity to examine—the material is calcium phosphate, and the structures are more like those of teeth of Sharks than of Lampreys, so that there can be no doubt that they are really teeth of fishes, and probably of fishes of somewhat higher grade than the Lampreys.[35] The Cambrian and Silurian specimens are said to be composed of calcium carbonate, which would render it more probable that, as has been suggested by Prof. Owen, they may have been teeth of some species of Sea-snail destitute of shell. It is, however, possible that they may have originally been horny, and that the animal matter has been replaced by carbonate of lime. Rohon and Zittel have recently shown that many of these are more allied to the teeth of worms than of any other animals.[36]

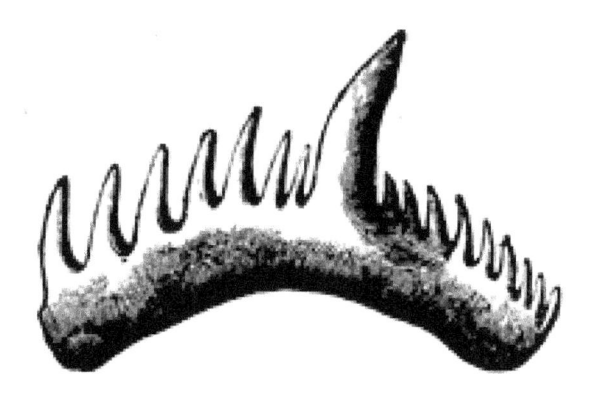

FIG. 103.—Lower Carboniferous Conodont. Magnified.—After Newberry.

If these older Conodonts were really teeth of fishes, they carry the introduction of these nearly as far back as that of the Mollusks and

Crustaceans. If they were not, then the earliest known representatives of this class belong to a much later age, that of the Silurian. Here we have undoubted remains of fishes belonging to two of the higher orders of the class; and in the succeeding Devonian these became multiplied and extended exceedingly.

Besides the inferior tribes already referred to, the modern seas and rivers present four leading types of fishes:—first, the ordinary bony fishes (Teleostians), such as the Cod, Salmon, and Herring; secondly, the Ganoid fishes, protected with bony plates on the skin, as the Bony-pike[37] and Sturgeon; thirdly, the Sharks and their allies, the Dog-fishes and Rays; fourthly, the peculiar and at present rare group of semi-reptilian fishes to which the name of *Dipnoi* has been given, on account of their capacity for breathing both in air and in water.

Of these four types the first is altogether modern, and includes the great majority of our present fishes. It does not make its appearance till the Cretaceous age, and then is at once represented by at least three of the modern families, those of the Salmon, Herring, and Perch. The history of the other three groups is precisely the opposite of this. They abound exceedingly at an early period, and dwindle to a much smaller number in the modern time. This is especially the case with the Ganoids and the Dipnoi. It is also remarkable that these groups of old-fashioned fishes[38] are in some respects the highest members of the class, approaching the nearest to the reptiles; but this accords with a well-known palæontological law, namely, that the higher members of low groups give way on the introduction of more elevated types, while the lower members may continue. Thus the decadence of these higher fish begins with the incoming of the reptiles, just as the decadence of the higher Mollusks and predaceous Crustaceans began with the incoming of the fishes. Further, the modern Ganoids and Dipnoi are mostly fresh-water animals, though the Sharks are largely pelagic. In the Palæozoic there seem to have been abundance of marine species of all these types; but though marine, they probably flourished most in bays and estuaries and on shallow banks; and the existence of these implies continental masses of land. This explains the curious coincidence that the introduction of fishes and of an abundant land flora synchronise, and that the ocean was still dominated by Invertebrates long after the fishes had become supreme in bays, estuaries, and rivers.

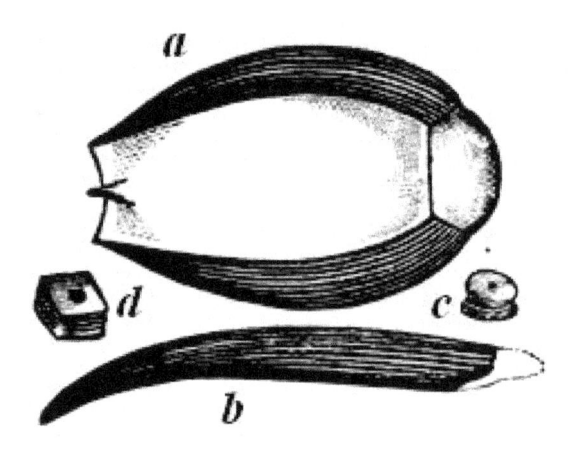

FIG. 104.—*a*, Head-shield of an Upper Silurian fish (*Cyathaspis*). *b*, Spine of a Silurian Shark (*Onchus tenui-striatus*, Agass.). *c, d*, Scales of *Thecodus*, enlarged.

The first fishes that we certainly know are the Ganoids and Sharks, which appear near the close of the Upper Silurian, in the English Ludlow for example (Fig. 104). The Ganoids found here all belong to an extinct group, characterised by the covering of the head and anterior part of the body with large bony plates. They are mostly small fishes, and probably fed at the bottom, and used their long or rounded bony snouts for grubbing in the mud for food. In this respect they present a singular resemblance to the Trilobites, so that we seem to have here animals of an entirely new type, the Vertebrate, and with bony instead of shelly coverings, taking up the *rôle* and, to some extent, the external form of a group about to pass away. Yet I presume that no derivationist would be hardy enough to affirm that the Trilobites could have been the ancestors of these fishes. Nor indeed is any ancestry even hypothetically known for them, for the doubtful Lampreys of the Cambrian Silurian are too remote and uncertain to be used in that way. The head-shield copied in outline in Fig. 104, and the restoration after Lankester in the frontispiece to this chapter, may serve to represent these curious primitive Ganoids, which are continued in the Devonian fishes represented in Figs. 105, 106.

FIG. 105.—*Cephalaspis Dawsoni* (Lankester). Lower Devonian of Gaspé.

Along with these, and not improbably their enemies, were certain Sharks (Fig. 104), known to us only by the spines which were attached to their fins as weapons of defence, and by detached bony tubercles which protected their skin. These remains are chiefly interesting as indications that two of the great leading divisions of the class of fishes originated together.

In the Devonian age the Ganoids and Sharks, thus introduced in the Silurian, may be said to culminate. The former, more especially, are represented by a great variety of species, some of them nearly allied to their Silurian predecessors (Fig. 106), others of forms and structure not dissimilar to those of the few surviving representatives of the order, or altogether peculiar to the Devonian (Fig. 107). So numerous are these fishes, and of so many genera and species—and this not merely in one region, but in widely separated parts of the world—that the Devonian has not inaptly been called the reign of Ganoids. As an illustration at once of the very peculiar forms of some of these fishes and of their wide distribution, I figure here along with the British species a *Cephalaspis* (Fig. 105) found in the Lower Devonian of Gaspé, in the same beds with some of the antique Devonian plants described in the last chapter.

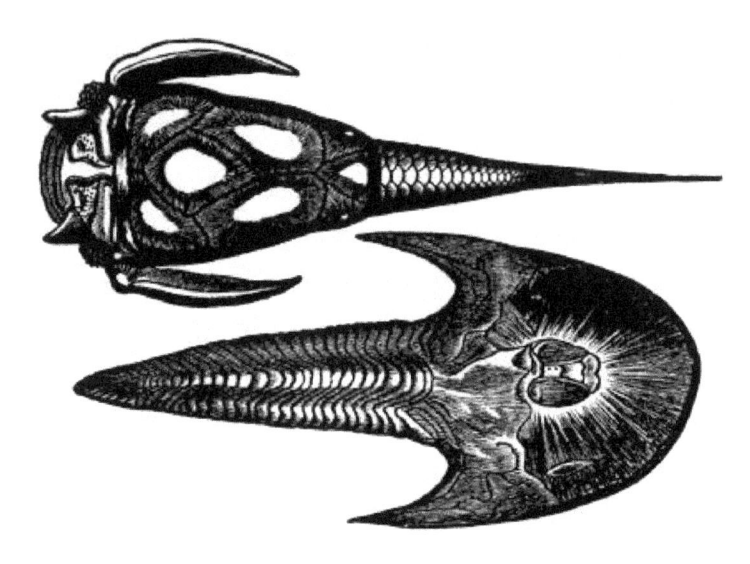

FIG. 106.—Devonian Placoganoid Fishes (*Pterichthys cornutus*, *Cephalaspis Lyelli*), from Scotland.

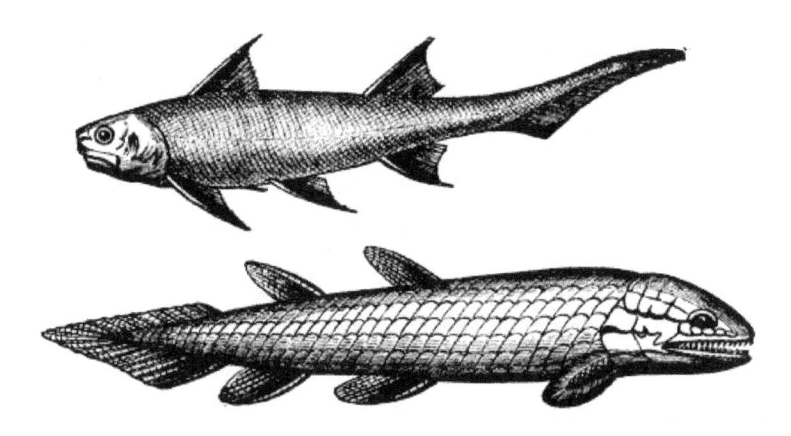

FIG. 107.—Devonian Lepidoganoid Fishes (*Diplacanthus* and *Osteolepis*). After Page and Nicholson.

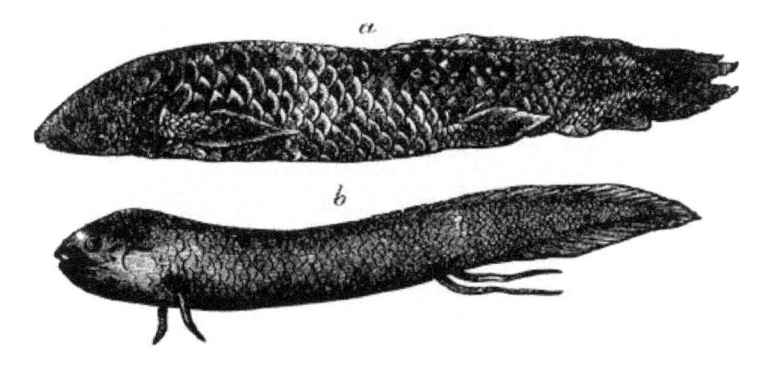

FIG. 108.—Modern Dipnoi.

a, Ceratodus Fosteri. Australia. *b, Lepidosiren annectus.* Africa.

A new and interesting light has recently been cast upon some of the most anomalous of the ancient fishes by the study of the now rare and peculiar species of the group of Dipnoi. Two of these, belonging to the genus *Lepidosiren*, are the "Mud-fishes" of the rivers of tropical Africa and America (Fig. 108, *b.*) These creatures have an elongated and elegant form, and the body is covered with overlapping horny scales like those of ordinary fishes; but the pectoral and ventral fins are rod-like, and are supported by simple cartilaginous rays, while the tailfin forms a fringe around the posterior part of the body. Unlike ordinary fishes, they have lungs as well as gills, and their mouths are armed with sharp, bony, beak-like teeth (Fig. 115), with which they can inflict terrible bites on the small fishes and frogs which furnish them with food. Their most remarkable habit is that of burying themselves in the mud of dried-up ponds, thus forming a sort of water-chamber or "cocoon," in which they remain in a torpid state until the return of the rainy season sets them free.

Another example of these Dipnoi is the Barramunda, or *Ceratodus* of the Australian rivers (Fig. 108*a*). This fish resembles the *Lepidosiren* in many essential points of structure; but its fins have lateral rays, and are consequently of some breadth, though of peculiar form, and its mouth is armed with flat, pavement-like teeth, wherewith it browses on aquatic grasses.

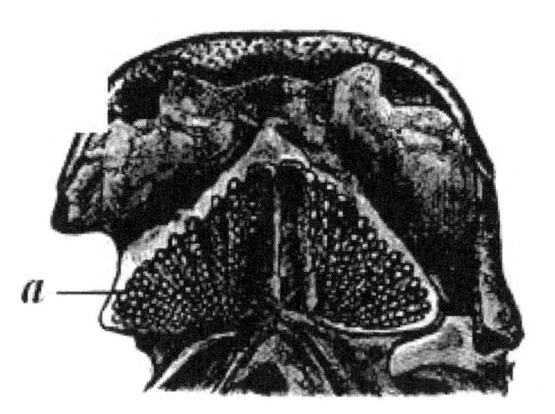

FIG. 109.—Anterior part of the palate of *Dipterus*. Showing the dental plates at *a*, Devonian.—After Traquair.

These modern fishes have enabled us to understand several mysterious forms met with in the older rocks. In the first place, they show the meaning of certain flat-toothed fishes, like *Dipterus* of the Devonian (Fig. 109), *Conchodus* of the Carboniferous (Fig. 110), and *Ceratodus* of the Carboniferous and Trias (Figs. 111, 112), previously of very doubtful character. These must all have been of similar structure and habits with the Barramunda, which is thus the sole survivor, perhaps itself verging on extinction, of a group of herbivorous fishes introduced, it may be, contemporaneously with the first stream affording the requisite vegetable food, and which have continued almost without improvement or deterioration to the present time. These fishes are, however, very closely connected with the Ganoids, and there are some of these, with fringed fins and overlapping scales, which, while regarded as true Ganoids, resemble the Dipnoi very closely.

FIG. 110.—Dental plate of *Conchodus plicatus* (Dn.). Coal-formation of Nova Scotia. *Acadian Geology*.

FIG. 111.—Dental plate of *Ceratodus Barrandii*. Coal-formation of Bohemia. After Fritsch.

FIG. 112.—Dental plate of *Ceratodus serratus*. From the Trias.

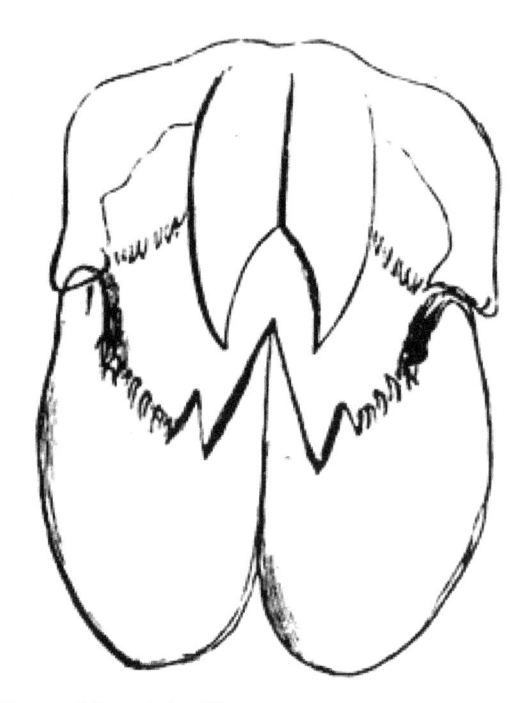

FIG. 113.—Jaws of *Dinichthys Hertzeri* (Newberry). Laterally compressed; one-sixth natural size.

Again, certain huge fishes, whose remains are found in the Devonian of Ohio,[39] had jaws on the same plan with those of *Lepidosiren*, but of enormous size and strength (Figs. 113, 114, 115), so that in this and some other points of structure they may be regarded as colossal Mud-fishes, and they must have had the same destructive powers, but on a far grander scale. They were besides clothed with heavy armour of bony scales, having some resemblance to that of those mailed fishes of smaller size already referred to, and indicating that, huge though they were, and formidable in destructive power, they also had enemies to be dreaded. These plates serve to ally them with the Ganoids, as their jaws do with *Lepidosiren*.

FIG. 114.—Lower Jaw of *Dinichthys Hertzeri*. One-sixth natural size.

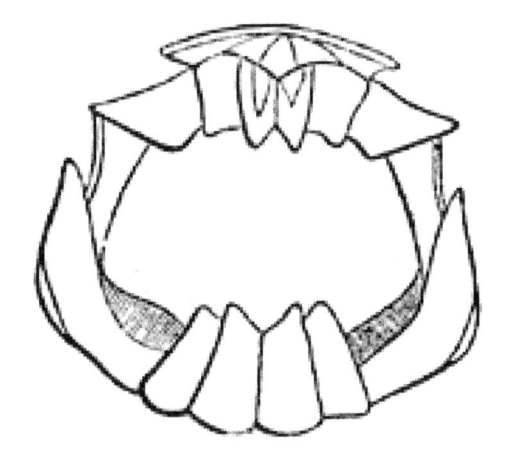

FIG. 115.—Jaws of *Lepidosiren*. Natural size.—After Newberry.

We are thus enabled to see in the streams, lakes, and bays of the Palæozoic, harmless fishes, of the type of *Ceratodus*, feeding on plants, and huge precursors of the Mud-fishes darting from the depths, and provided with a dental apparatus more formidable than that of any modern fish, sufficient to pierce the strongest armour of the Ganoids, and to destroy and devour the largest aquatic animals. These huge fishes, armed with shears two or three feet in length, and capable of cutting asunder scale, flesh, and bone, are the *beau idéal* of destructive monsters of the deep, far surpassing our modern Sharks; and if, by means of supplementary lungs, they could breathe in air as well as in water, they would on that account be all the more vigorous and voracious.

Newberry has well remarked that while in the Devonian the Ganoids and Dipnoi were the real tyrants of the sea, as well as of the streams, in the Carboniferous they already diminish in size, though still abundant as to numbers, and are more limited to estuaries and fresh waters. Thus their departure from power had already begun, and went on until in modern times the proportion of Ganoids to ordinary fishes is, according to Günther, nine out of 9,000. The Carboniferous, indeed, very specially abounds in small Ganoids, though there are many large and formidable species. One of these smaller species, a very beautiful little fish, of fresh-water ponds and streams in the older part of the Carboniferous age, is represented of the natural size in Fig. 116, and is not a restoration, being found preserved entire, though flattened, in a fine bituminous shale, which has perfectly preserved even the most delicate sculpturing of its bony scales.

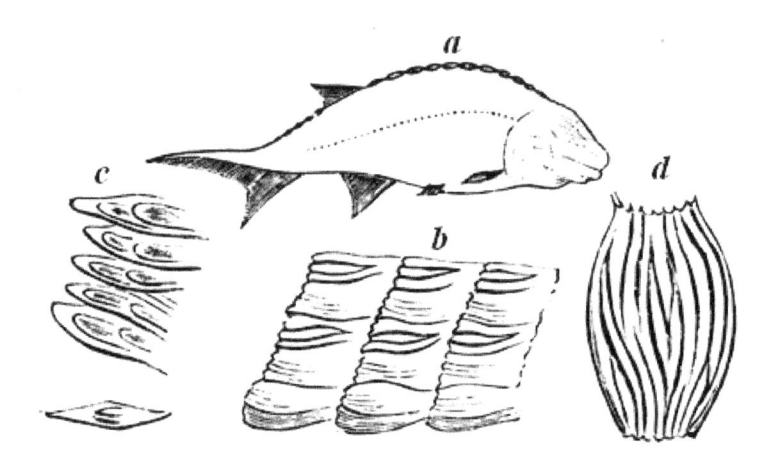

FIG. 116.—A small Carboniferous Ganoid (*Palæoniscus (Rhadinichthys) Modulus* Dn.). Lower Carboniferous, New Brunswick.

a, Outline. *b, c, d*, Sculpture of scales magnified.

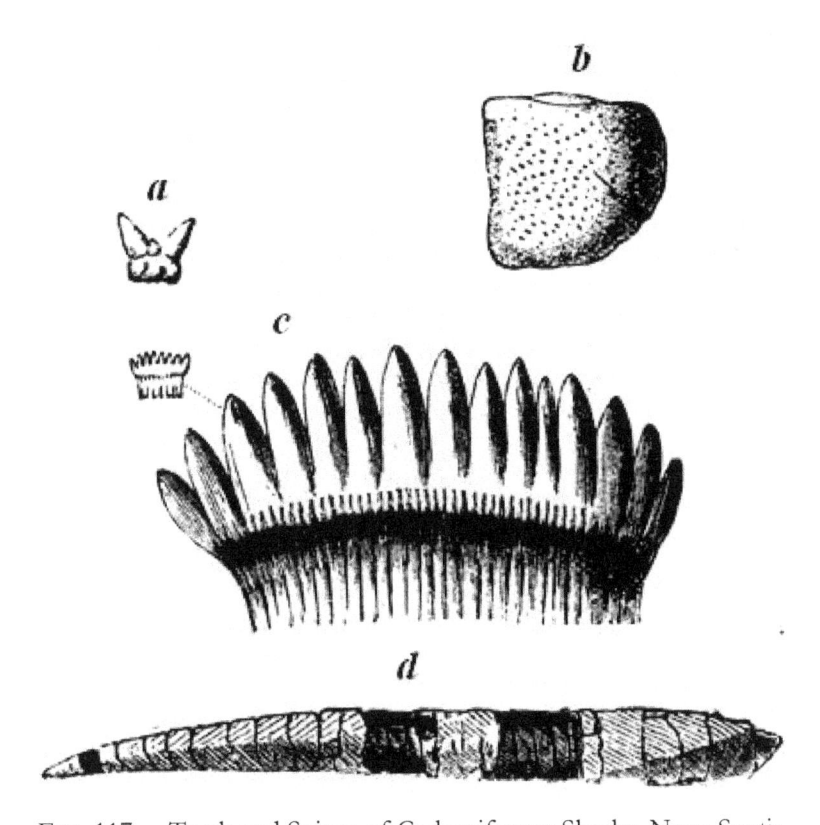

FIG. 117.—Teeth and Spines of Carboniferous Sharks. Nova Scotia.

a, *Diplodus penetrans*. *b*, *Psammodus*. *c*, *Ctenoptychius cristatus*. *d*, Spine, *Gyracanthus magnificus*. One-eighth natural size.—*Acadian Geology*.

The Sharks in the Carboniferous increase in number and importance. Fig. 117 shows a few examples of their teeth and spines. In the Carboniferous, however, there is a great preponderance of those species with flat, crushing teeth fitted for grinding shells,[40] which in diminishing numbers continue up to the present time, when they are represented by the Port Jackson Shark and a few other species. The increase toward the modern time of the true Sharks[41] with sharp cutting teeth, is obviously related to the increase of the ordinary fishes which furnish them with food. Another curious difference, connected probably with the same circumstance, is the fact that in the sharp toothed Sharks of the Carboniferous the two side fangs of each tooth are the largest, or are exclusively developed (Fig. 117, *a*), while in later periods the central point becomes dominant, or is developed to the exclusion of the others (Figs. 118, 119).

    The Ganoids and Dipnoi still, however, occupy a very important place through the Mesozoic ages (Fig. 120), and it is only at the close of the

Cretaceous that they finally give place to the Teleosts, or common fishes, which, though perhaps more fully specialised in purely ichthyic features,

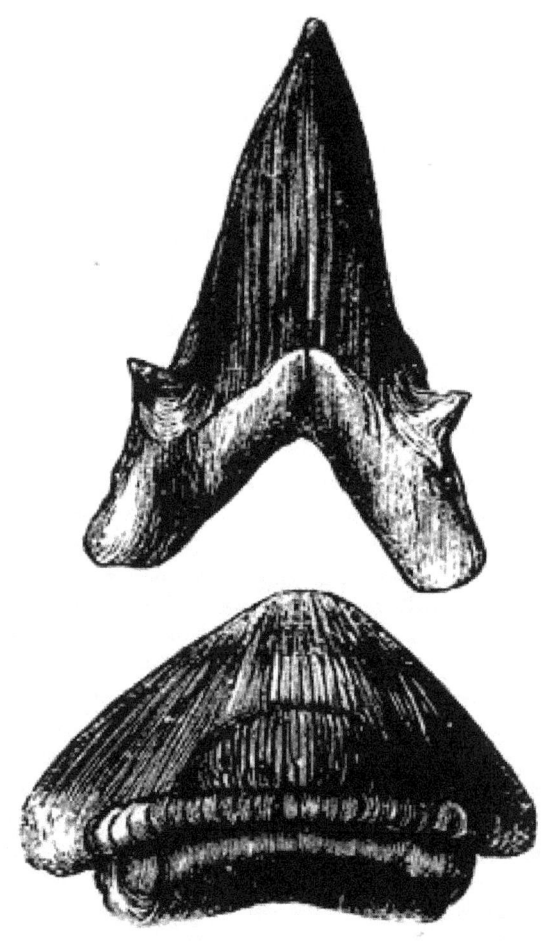

have dropped the reptilian characteristics of

FIG. 118.—Teeth of Cretaceous Sharks (*Otodus* and *Ptychodus*).—After Leidy. their predecessors (Fig. 121). It is interesting to observe that these old-fashioned fishes had culminated before the advent of air-breathing Vertebrates, which appear for the first time in the Carboniferous. It is further to be observed that groups of fishes furnished with means of aiding their gills by rudimentary lungs were especially suited to waters more charged with carbonic acid, and less with free oxygen, than those of more recent times. This remark especially applies to the mephitic and sluggish streams and lagoons of the Carboniferous swamps, where, in the midst of a rank vegetation and reeking masses of decaying organic matter, the half air-breathing fishes and the amphibious reptilian animals met with each other and found equally congenial abodes. Thus, independently of the fact that some of these fishes were probably vegetable feeders, it is not altogether an accident, but a wise adaptation, that caused the culmination of the reptilian fishes and batrachian reptiles to coincide with the enormous development of the lower forms of land-plants in the Devonian and Carboniferous. Another curious illustration of the diminishing necessity for air-breathing to the fishes, is the change of the tail from the unequally-lobed

FIG. 119.—Tooth of a Tertiary Shark (*Carcharodon*). or heterocercal form, which prevailed in the Palæozoic, to the more modern equally-lobed (homocercal) style in the Mesozoic. The former is better suited to animals which have to rise rapidly to the surface for air, and is still continued in some modern fishes, which for other reasons need to ascend and descend, or to turn themselves in the water; but the homocercal form is best suited to the ordinary fish, whether Ganoids or Teleosts (Fig. 122). It is curious also to find the beginning of the dominancy of the ordinary fish to coincide with that of the broad-leaved exogenous trees in the later Cretaceous, and to precede immediately the appearance of the mammals on the land; all these changes being related to the purer air, the clearer waters, and the more varied continental profiles of the later geological periods. Thus physical improvement and the changes of animal and vegetable life are linked together by correlations which imply not only design, but prescience, whether we attribute these qualities to a spiritual Creator or to mere atoms and forces.

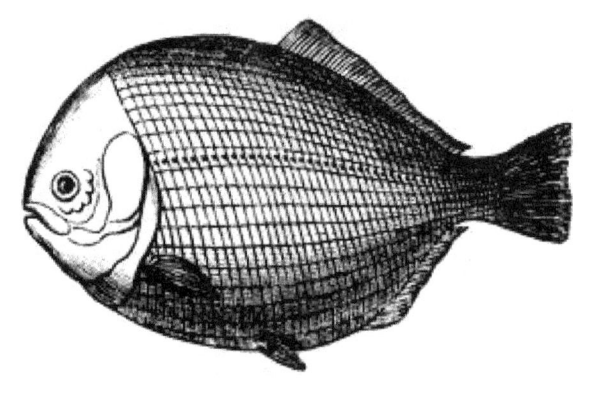

FIG. 120.—A Liassic Ganoid (*Dapedius*). Restored.—After Nicholson.

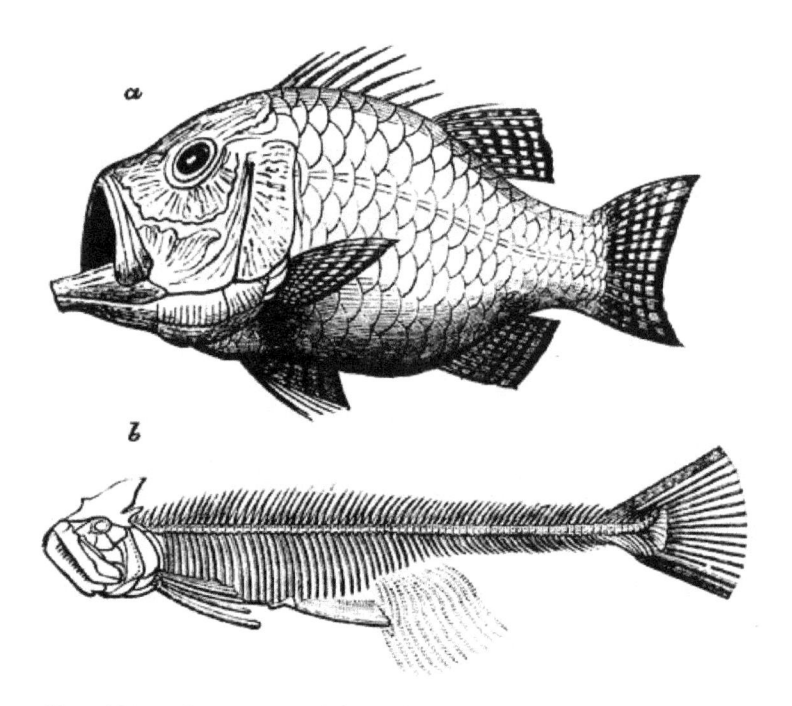

FIG. 121.—Cretaceous Fishes of the modern or Teleostian type.

a, *Beryx Lewesiensis*. English chalk. *b*, *Portheus molossus* (Cope). A large fish from the American Cretaceous. One twenty-eighth natural size.

The history of fishes extends further through geological time than that of any other Vertebrates, and is perhaps more completely known to us, in consequence of the greater facilities for the preservation of their remains in aqueous deposits. If we receive Pander's Conodonts as indicating a low type of cartilaginous fishes, these must have continued for vast ages without any elevation, and struggling for a bare existence amidst formidable Cuttle-fishes and Crustaceans, before, under more favourable conditions, they suddenly expanded into the high and perfect types of Ganoids and Sharks. If we reject the early Conodonts, then the two last-mentioned types spring together and suddenly into existence, like the armed men from the dragon's teeth of Cadmus. They rapidly attain to numbers and grandeur unexampled in later times, and become the lords of the waters at the time when there was probably no Vertebrate life on the land. As the reptiles establish themselves on the land and in the waters, the Ganoids diminish, but the Sharks hold their own. At length the reign of reptiles is over, but the Ganoids, instead of resuming their pristine numbers, give place to the Teleosts, and become reduced to insignificance; while the Sharks, profiting by the decadence of the great marine reptiles, remain the tyrants of the seas. This history is strangely unlike a continuous evolution; but we are anticipating facts which will fall to be discussed in a subsequent chapter.

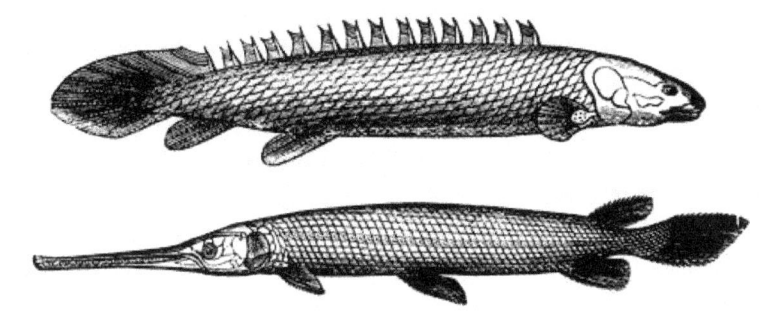

FIG. 122.—Modern Ganoids (*Polypterus*. Africa. *Lepidosteus*. America).

A Microsaurian of the Carboniferous Period (*Hylonomus Lyelli*). Restored from the skeleton and dermal appendages found in an erect Sigillaria. Half natural size.

CHAPTER VI.

THE FIRST AIR-BREATHERS.

WERE our experience limited to the animals whose remains are found in the earlier Palæozoic rocks, we might be unable to conceive the possibility of an animal capable of living and breathing in the thin and apparently uncongenial medium of air. More especially would this appear doubtful if our experience of the atmosphere presented it to us as loaded with carbonic acid, and less rich in vital air than it is at present. Even the mechanical difficulties of the case might strike us as considerable, in our ignorance of the capabilities of limbs. Still, as time wore on, we should find this problem worked out along three distinct lines of advancement—those of the Mollusk, the Arthropod, and the Vertebrate, and in each of these with different machinery, related to the previous locomotive and water-breathing apparatus of the type.

Respiration under water depends, not on the water itself, but on the small percentage of free oxygen which it contains, and this is utilised for the aëration of the blood of animals, by that wonderful and often extremely beautiful apparatus of delicate fibres or laminæ penetrated with blood-vessels, which we call a gill. Except those lowest creatures which aërate their blood merely at the general surface of the body, all animals capable of respiration in water are provided with gills in some form, though in many of the humbler types, like that of the familiar Oyster, the gills are used for the double purpose of aërating the blood and, by their minute vibrating threads or cilia, drifting food to the mouth.

In the great group of radiated animals, the *Protozoa*, *Cœlenterata*, and *Echinodermata*, no air-breathing creature exists, or, in so far as is known, has existed, so that this vast group of animals is limited altogether to the waters; and this is undoubtedly one mark of its inferiority.

In the sub-kingdom of the Mollusks the highest class, that of the Cuttle-fishes and Nautili, has been, singularly enough, rejected as unfit for this promotion, though it was early introduced, and attains to a high development of muscular energy and nervous power. The group next in order, that of the Snails and their allies, alone ventures in some of its families to assume the *rôle* of air-breathing. As might be expected, in creatures of this stamp the simplest means are employed to effect the result. In the sub-aquatic species the gills are contained in a chamber, where they are protected and kept supplied with water. In the air-breathing species, this gill-chamber is merely emptied of its contents and converted into an air-sac or functional lung. Thus a rude and imperfect method of air-breathing is contrived, which scarcely separates the animals that possess it from their aquatic relatives, but which nevertheless gives to us the beautiful and varied groups of the Land-snails and of the air-breathing fresh-water Snails.

In the worms and Crustaceans the gills are placed at the sides of the body, and connected with its several segments. But the Crustaceans, like the Cuttle-fishes, though the highest aquatic type, never become air-breathers. It is true some of them, like the Land-crabs, live in the air; but they retain their gills, and have to carry with them a supply of water to keep these moist.

But in order to elevate the Annulose type to the true dignity of air breathing, three new classes had to be introduced, differing altogether in their details of structure; and all three seem to have been placed on the earth about the same time. They are: First, the Myriapods, or Gallyworms and Centipedes; secondly, the Insects; and thirdly, the Arachnidans, or Spiders and Scorpions.

In the Myriapods a system of air-tubes, kept open by elastic spiral fibres, penetrates the body by lateral pores, thus retaining the resemblance to the lateral respiration of the Crustaceans and worms. In the Insects, where this type of structure rises to its highest mechanical perfection, and where the animal is enabled to be not merely an air-breather, but a flier, the same system of lateral pores and internal air-tubes is adopted, and is so extended and ramified as to give a very perfect respiration. In the Spiders and Scorpions the system is the same, except that in the latter and a part of the former the whole or a part of the tracheal system becomes expanded into air-chambers simulating true lungs.

Among the Vertebrates, the fishes are breathers by gills attached to arches at the sides of the neck. But already in the Devonian we have reason to believe that there were fishes having the swimming-bladder opening into the back of the mouth to receive air, and divided into chambers, so as to constitute an imperfect lung. And here we have not, as in the lower types, an adaptation of the old water-breathing organs, but an entirely new apparatus. In the next grade of Vertebrates we find, as in the Frogs, Water-lizards, etc., that the young are aquatic and breathe by gills, while the adults acquire lungs, sometimes retaining their gills also, but in the higher forms parting with them. Thus in the vertebrates alone we have true lungs, distinct structurally from gills; and these lungs attain to their highest perfection in the birds and mammals.

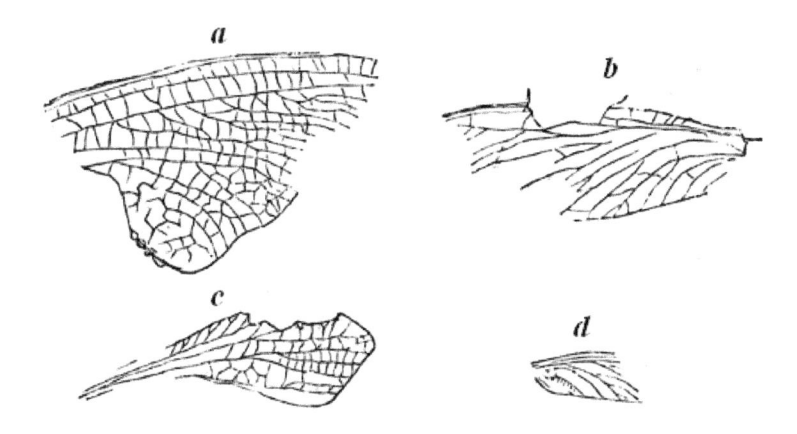

FIG. 123.—Wings of Devonian Insects. Middle Devonian of New Brunswick.

a, Platephemera antiqua (Scudder). *b, Homothetus fossilis* (Scudder). *c, Lithentomum Harttii* (Scudder). *d, Xenoneura antiquorum* (Scudder).

The oldest air breathers at present known are Scorpions and insects allied to the modern May-flies, which have been found in the Silurian. Next to these, and more important in number and variety, are the insects of the Erian plant beds of New Brunswick. They were discovered by the late lamented Prof. C. F. Hartt in the plant-bearing shales of the Middle Devonian (Fig. 123). The beds containing them hold also a species of *Eurypterus*, an obscure Trilobite, and a Crustacean allied to the modern Stomapods,[42] besides a shell which may possibly be that of a Land-snail, to be mentioned in the sequel. They are also exceedingly rich in beautifully-preserved remains of Devonian plants. The collection made by Prof. Hartt is limited to a few fragments of wings; but these, in the skilful hands of Mr. Scudder, have proved to be rich in geological interest. One is a gigantic *Ephemera* or May-fly, which must have been five inches in the expanse of the wings, which are more complex in their venation than those of its modern allies (Fig. 123, *a*). Another presents peculiarities between those of the May-flies and Dragon-flies (Fig. 123, *b*). A third is a Neuropter, not belonging to any known family, but allied to some in the Coal-formation (Fig. 123, *c*). A fourth (Fig. 123, *d*) is a small and delicate wing, supposed to have belonged to an animal having some points of resemblance to the modern crickets. Two others are represented by mere fragments of wings, insufficient to determine their affinities with certainty. No other insects of this age have been discovered elsewhere; but it is to be borne in mind that no other locality rich in Devonian plants has probably been so thoroughly explored. The hard slaty ridges containing these fossils are well exposed on the coast near the city of St. John, and Messrs. Hartt and

Matthew of that city, acting, I believe, in concert with and aided by the Natural History Society of the place, not only searched superficially, but removed by blasting large portions of the richest beds, and examined every fragment with the greatest care. Their primary object was fossil plants, of which they obtained magnificent collections; and it is scarcely possible that the insects could have been found but for the exhaustive methods of exploration employed.

It is interesting to observe, respecting these oldest insects, that they all belong to those families which have jaws, and not suctorial apparatus, that they are not of those which undergo a complete metamorphosis, and that their modern congeners pass their larval stage in the water. Thus the waters gave birth to the first insects, and their earliest families were not of those which suck honied juices or the blood of animals, or which pass through a worm-like infancy. These groups belong apparently to much later times.

On one of the specimens collected by Messrs. Hartt and Matthew, and placed by them in my hands, is a spiral form which in every particular of external marking resembles a genus of modern West Indian Land-snails.[43] I have hesitated to describe it, as the structure is lost and the form imperfect; but I cannot help regarding it as an indication that this group of land animals also will be traced back to the Devonian age.

Ascending from the Devonian to the Carboniferous, we at once find ourselves in the midst of air-breathers of various types. Here are Myriapods, insects of several orders, Spiders, Scorpions, Land-snails, and Batrachian reptiles, and these of many species, and found in many localities widely separated. We can thus people those dark, luxuriant forests, to which we owe our most valuable beds of coal, with many forms of life; and as most of these belong to tribes likely to multiply abundantly where food was plentiful, we can imagine multitudes of Snails and Millepedes feeding on succulent or decaying vegetable matter, swarms of insects flitting through the air in the sunnier spots, while their larvæ luxuriated in decaying masses of leaves or wood, or peopled the pools and streams. In like manner, in imagination we can render these old woods vocal with the trill of crickets and with the piping or booming of smaller and larger Batrachians. Let us now, in accordance with our plan, inquire as to the nature of these early air-breathers and the fortunes of their families in the geological history.

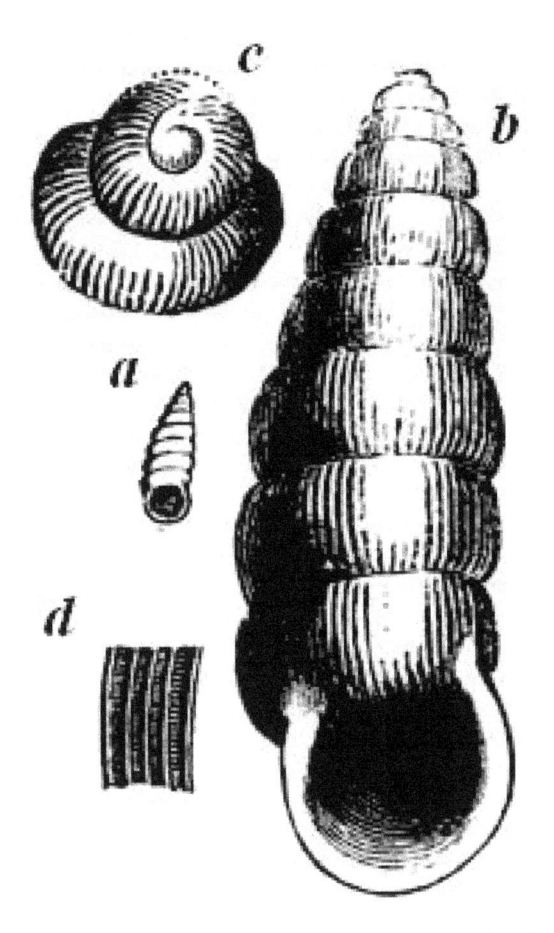

FIG. 124.—Land-snail (*Pupa vetusta*, Dawson). From the Coal-formation. *a*, Natural size. *b*, Magnified. *c*, Apex. *d*, Sculpture. Enlarged.

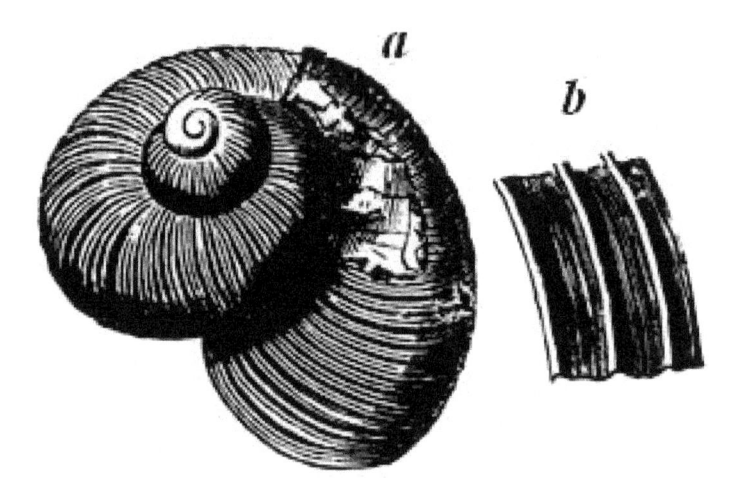

FIG. 125.—Land-snail (*Zonites (Conulus) priscus*, Carpenter). From the Coal-formation.

a, Shell. Enlarged; the line below shows the natural size. *b*, Sculpture. Enlarged.

The Land-snails known as yet in the Carboniferous are limited to five or six species, belonging to four genera, all American and related to existing American forms. The two earliest known are represented in Figs. 124 and 125.[44] One of them is a *Pupa*, or elongated Land-snail, so similar to modern forms that it does not merit a generic distinction, and is indeed very near to some existing West Indian species. The other is in like manner a member of the modern genus *Zonites*. These are from the Coal-formation of Nova Scotia, and the Pupa must have been very abundant, as it has been found in considerable numbers in a layer of shale, and in the stumps of erect trees, in beds separated from each other by a thickness of 2,000 feet of strata. The *Zonites* is much more rare. A second Pupa is found in Nova Scotia, and two species occur in the Coal-field of Illinois. One of these is a Pupa still smaller than *P. vetusta*, and, like some modern species, with a tooth-like process on the inner lip.[45] The other has been placed in a new genus, but is very near to some of the smaller American Snails still living. Its most special character is a plate extending from the inner lip over half the aperture, a contrivance for protection still seen in some modern forms. Thus the Land-snails come on the stage in at least three generic forms, similar to those which still live, but all of small size, indicating perhaps that the conditions were less favourable for such creatures than those of the temperate and warmer climates at present. It may seem a small step in advance for Sea-snails to lose their gills and to become Land-snails, and this without any elevation of their general structure; but it must be borne in mind that we have here not only the dropping of the gills for an air-sac, but profound changes in teeth, mucous glands, shell, and other particulars, to fit them for new food and new habits. It is also singular that the Land-snails at once appear instead of the intermediate forms of the air-breathing fresh-water snails. These last may, however, yet be found.

The Millepedes, like the Land-snails, were first found in the Coal-formation of Nova Scotia, but species have since been discovered not only in Illinois, but also in Great Britain and in Bohemia. In Nova Scotia alone two genera and five distinct species have been found, all in the interior of erect trees, to which these creatures probably resorted for food and shelter (Fig. 126). All the species yet known are allied to the modern Gallyworms, though presenting special features which seem to separate them as a distinct family,[46] and were probably vegetable-feeders. Some of the species have the peculiarity, unknown among their modern successors, of being armed with long spines.[47] The moist, equable climate and exuberant vegetation of the Coal-period would naturally be very favourable to Millepedes, and it is likely that the discoveries made as yet give but a faint idea of their actual abundance. It is not improbable that they subsequently declined, as we know of none between the Carboniferous and the Jurassic, and they do not seem to have improved up to the modern period. The Carnivorous Myriapods, however, or Centipedes proper, a higher and essentially distinct type, are not known until much more recent times.

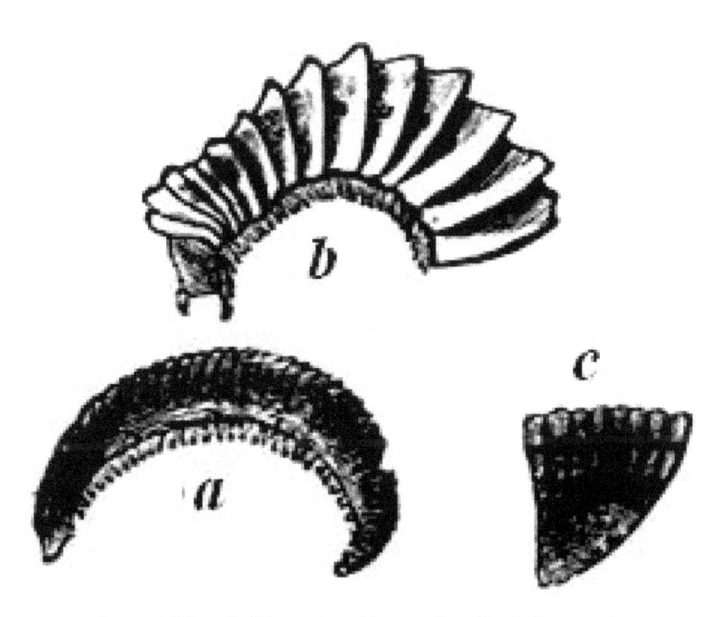

FIG. 126.—Millepedes. From the Coal-formation.

a, *Xylobius sigillariæ* (Dawson). *b*, *Archiulus xylobioides* (Scudder). Anterior segments. Enlarged, *c*, *X. farctus* (Scudder). Caudal portion. Enlarged.

The insects of the Carboniferous as yet known, belong to three out of the ten or more orders into which the class is divided. One of these is represented by a number of species of Cockroach, another by May-flies and a Dragon-fly, and another by some weevil-like Beetles. The Cockroach is characterised by Huxley as one of the "oldest, least modified, and in many ways most instructive forms of insects;" and both he and Rolleston take its anatomy as typical of that of the class. That these creatures should have abounded in the Coal-period we need not wonder, when we consider the habits of those that infest our houses, and when we further bear in mind the number of species, some of them two inches in length, that exist in tropical climates. So many species of this family have been found in the Coal-formation on both sides of the Atlantic,[48] that we may fairly regard them as constituting one of its most characteristic features, and as probably the oldest representatives of the order to which they belong[49] (Fig. 127). There were also in the Coal-period insects allied to the Locusts and to the Mantids, a carnivorous group. One of the latter (*Lithomantis*), described by Woodward, is a magnificent insect, not unlike some modern tropical species. It was found in the Coal-formation of Scotland. A still larger species, probably the largest insect known, has been described by Brongniart. The May-flies (*Ephemeridæ*) are represented in the Carboniferous by several very large species. That of which the wing is shown in Fig. 128 must have been seven inches in expanse of wings. The habits of the modern May-flies show us how animals of this group, living as larvæ in the streams and lakes, must have afforded large supplies of food to fishes, and when mature must have emerged from the waters in countless myriads, filling the air for the brief term of their existence in the perfect state. The May-flies represent another insect order.[50] The Coal-measures of Saarbruck have afforded several species allied to the white ants (*Termites*), insects which must have found abundant scope for their activity in the dead trees of the carboniferous forests. The occurrence of beetles,[51] especially of the weevil family, which have as yet been found only in Europe, might have been expected, considering the habits and modern distribution of this group. It has been asserted that moths[52] have been found in the Carboniferous; but the proof of this, so far as known to me, is the occurrence of leaves, noticed by Sternberg, with markings similar to those made by the larvæ of minute leaf-mining moths. This, however, is uncertain evidence. If we consider the orders of insects not found in the Coal-formation, we can perceive good reasons for the absence of some of them. Those containing the lice and fleas, and other minute and parasitic insects, we can scarcely expect to find. The bees and wasps, and the butterflies and moths, are little likely to have been present where there were scarcely any flowering plants; but such groups as those of the two-winged flies, the plant-bugs and the ants, we might have expected, but for the fact of their being highly specialised forms, and for that reason likely to have appeared later.[53] There are, indeed, as yet no haustellate or suctorial insects known in this early period. Plausible theories of the phylogeny of insects are not wanting; but they do not well suit the known facts as to their first appearance; and perhaps we may venture without much blame to apply to the insects of the Coal-period the remark made by Wollaston with reference to the rich insect fauna of the isolated rock of St. Helena: "To a mind which, like my own, can accept the doctrine of creative acts as not necessarily 'unphilosophical,' the mysteries [of the existence of these species in an island so remote from other lands], however great, become at least conceivable; but those which are not able to do this may, perhaps, succeed in elaborating some special theory of their own, which, even if it does not satisfy all the requirements of the problem, may at least prove convincing to themselves."

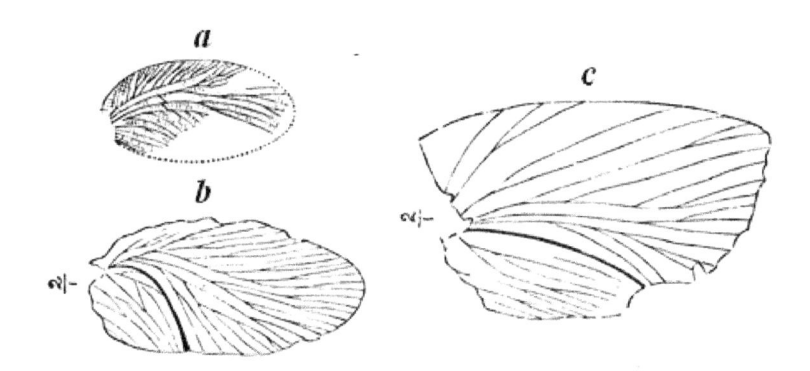

FIG. 127.—Wings of Cockroaches. From the Coal-formation.

a, Archimulacris Acadicus (Scudder). *b, Blattina Bretonensis* (Scudder). *c, B. Hesri* (Scudder).

FIG. 128.—Wing of May-fly (*Haplophlebium Barnesii*, Scudder). From the Coal-formation.

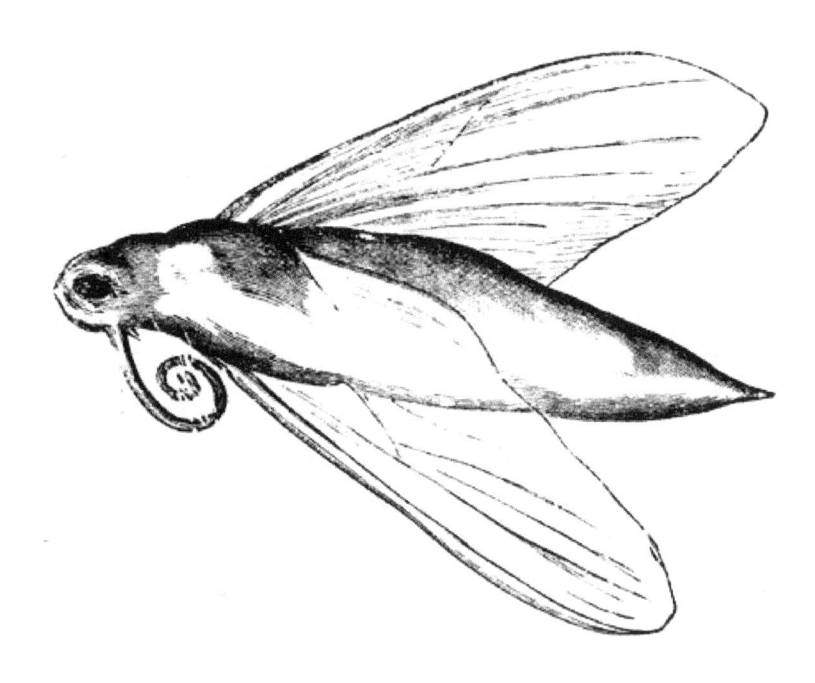

FIG. 129.—A Jurassic Sphinx-moth (*Sphinx Snelleri*, Weyenburgh).

FIG. 130.—An Eocene Butterfly (*Prodryas persephone*, Scudder). From Colorado.

The suctorial insects make their first certain appearance in the Jurassic; and the magnificent Sphinx Moth in Fig. 129 is an example of the magnitude and perfection to which that tribe attained in the age of the Solenhofen slate; though Weyenburgh, who describes it, fancies that he sees evidence that it may, unlike any modern moths, have been provided with a sting. The most perfect and beautiful fossil butterfly known to me is that represented in Fig. 130, from a photograph kindly given to me by Mr. Scudder. It is from the Tertiary rocks of Western America, and is laid out in stone as neatly as if prepared by an entomologist, while its preservation is so perfect that even the microscopic scales on the wings can be made out. It belongs to one of the highest types of modern butterflies, that to which the *Vanessæ* belong, but with some points of structure pointing to the lower group of the "Skippers" (*Hesperiadæ*). Scudder remarks that while the fore-wings resemble those of the former group, the hind-wings look more like those of the latter; and this seems to be a common character of two or three others of the few fossil species known, none of which are older than the Tertiary.

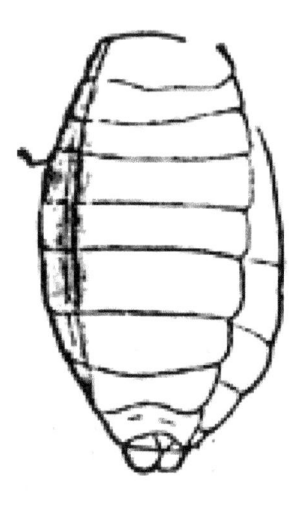

FIG. 131.—Abdominal part of a Carboniferous Scorpion.[54]

We know too little of the spiders and scorpions of the Carboniferous to say more than that they closely resemble modern forms. Two of the scorpions are represented in Figs. 131 and 132; and the only spider certainly known, which is from Silesia, is said to belong to the group of the hunting or trap-door spiders (*Lycosa*).[55]

The Batrachians of the Coal are its most characteristic and remarkable air-breathers,—especially so as the precursors of the reptiles of the Mesozoic age. Cope in a recent summary enumerates no less than thirty-nine genera and about one hundred species; and to these have to be added at least a dozen more recently discovered in Europe; though it was only in 1841 that the first indications of such creatures were found, and were then regarded by geologists with the same scepticism which some of them still apply to *Eozoon*. The first trace ever observed of batrachians in the Carboniferous consisted of a series of small but well-marked footprints found by the late Sir W. E. Logan in the Lower Carboniferous shales of Horton Bluff, in Nova Scotia. In that year this painstaking geologist had examined the coal-fields of Pennsylvania and Nova Scotia, with the view of following up his important discovery of the *Stigmariæ*, or roots of *Sigillaria*, as accompaniments of the coal-underclays. On his return he read a paper, detailing his observations, before the Geological Society of London. In this he mentioned the footprints in question; but the paper was published only in abstract, and the importance of the discovery was overlooked for a time, the anatomists evidently being shy to acknowledge the validity of the evidence for a fact so unexpected. Fig. 133 is a representation of another slab subsequently found in beds of the same age in Nova Scotia, and which may serve to indicate the nature of Sir William's discovery. In consequence of the neglect of this first hint by the

London geologists, the discovery of bones of a batrachian by von Dechen at Saarbruck in 1844, and that of footprints by King in Pennsylvania in the same year, are usually represented as the first facts of this kind. My own earliest discovery of reptilian bones in Nova Scotia was made in 1844, though not published till some time afterward, and was followed up by further collections in company with Sir Charles Lyell in 1851, at which time also the earliest land-snail was found, and in the following year the first millepede. Since that time the progress of discovery has been astonishingly rapid, and has extended over most of the principal coal-areas on both sides of the Atlantic.

FIG. 132.—Carboniferous Scorpion (*Eoscorpius carbonarius*, Meek and Worthen). Illinois.

FIG. 133.—Footprints of one of the oldest known Batrachians, probably a species of *Dendrerpeton*. From the Lower Carboniferous of Parrsboro, Nova Scotia. Upper figure natural size.

We may, for convenience, call these animals reptiles, but they are regarded as belonging to that lower grade of reptilian animals, the Amphibians or Batrachians, which includes the modern frogs and newts and water-lizards.[56] Still it would be doing great injustice to the carboniferous reptiles not to say, that while related to this low type, they presented a much greater range of organisation than it shows at present, evincing a capability to fill most of the places now occupied by the true reptiles. Some of them were aquatic, and with limbs rudimentary or little developed, but many of them walked on the land, and were powerful and predaceous creatures. They had large and complex teeth, they were protected by external bony plates, and some of them had in addition a beautiful covering of horny plates and spines, and ornamental lappets. Many had well-developed ribs, indicating a condition of respiration much in advance of that in the ribless batrachians. Some of them attained to size and strength rivalling those of the modern alligators, while some of the smallest species exhibit characters approaching in some respects to the lizards.

Perhaps the most fish-like of these animals are those first discovered by von Dechen (*Archegosaurus*, Fig. 134). Their long heads, short necks, supports for gills, feeble limbs and long flat tail, show that they were aquatic creatures presenting many points of resemblance to the Ganoid fishes which must have been their companions. Yet they show what no fish can exhibit, fore and hind limbs with proper toes, and the complete series of bones that appear in our own arms and legs, while they must have had true lungs and breathed through nostrils. So different are they from the fish in details, that a single limb bone, a vertebra, a rib, or a fragment of a skull bone, suffices to distinguish them. Much has been said recently of the genesis of limbs; and here, as far as now known, we have the first true limbs; but it is scarcely too much to say that the feet of *Archegosaurus* differ more from the fins of any carboniferous fish than they do from the human hand; while it is certain that the feet which made the impressions represented in Fig. 133, on the lowest beds of the Carboniferous, or that from the upper coal-formation represented in Fig. 139, were not less typical or perfectly formed feet than those of modern lizards.

Leaving these fish-like forms, we find the remainder of the carboniferous reptiles to diverge from them along three lines.

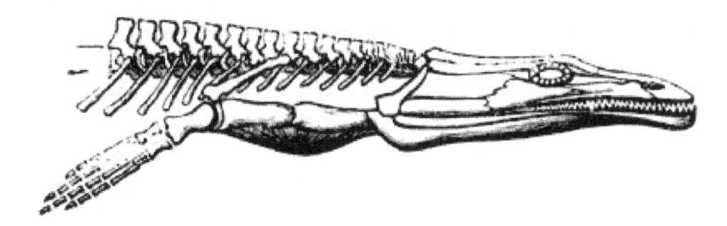

FIG. 134.—*Archegosaurus Decheni*. Head and anterior limb reduced. Coalfield of Saarbruck.

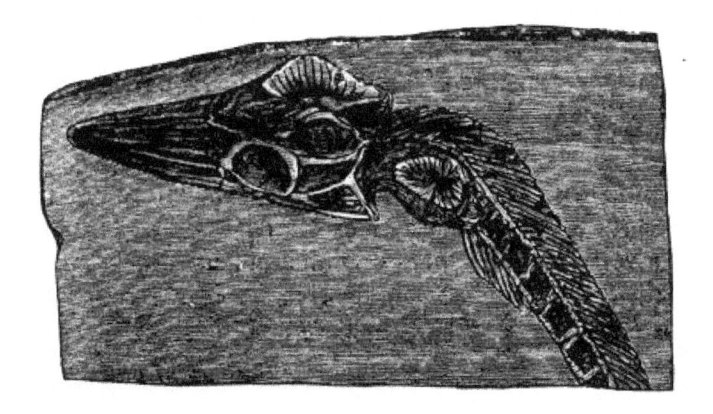

FIG. 135.—*Ptyonius*. A Snake-like Amphibian. Coal-measures of Ohio.—After Cope.

The first leads to snake-like creatures, destitute of limbs, and which must have been functionally the representatives of the serpents in the Palæozoic, though batrachian in their affinities (Fig. 135). They are found both in Europe and America; and Huxley describes one from Ireland more than twenty-one inches long, and with over one hundred vertebræ.[57] Some extraordinary traces are found on the sandstones of the coal-formation,[58] which appear to indicate that there may have been species of this type much larger than any represented by skeletons, and with bodies perhaps six inches in diameter. It is not unlikely that they had the habits of the modern water snakes.

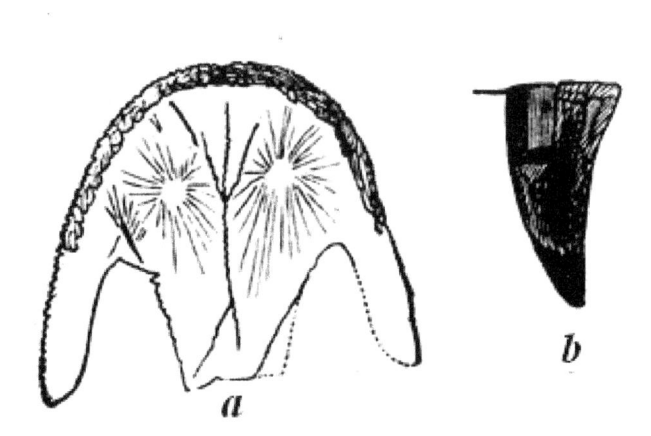

FIG. 136.—A large Carboniferous Labyrinthodont (*Baphetes planiceps*, Owen).

a, Anterior part of the skull, viewed from beneath. One-sixth natural size, *b*, One of the largest teeth, natural size.

A second line leads upward to large crocodile-like creatures, with formidable teeth, strong bony armour, and well-developed limbs (*Labyrinthodontia*, Figs. 136, 137). Some of them must have attained a length of ten feet. They were lizard-like in form, could walk well, as is seen from the footprints of some of the species which present a considerable stride, and moved over mud without the belly touching the ground. Their tails were long, and probably useful in swimming. Their heads were flat and massive, and their teeth were strengthened by a remarkable folding inward of the outer plate of enamel (Fig. 137 *b*). The belly was protected by bony plates and closely imbricated scales. In some of the species at least the upper parts were clothed with horny scales, and the throat and sides were ornamented with pendent scaly fringes or lappets.[59] Their general aspect and mode of life must have resembled those of modern alligators; and in the vast swamps of the Coal-period, full of ponds and sluggish streams swarming with fish, they must have found a most suitable abode. While rigid anatomy may ally these animals rather with the batrachians than the true reptiles, it is evident that their great size, their capacity for walking with the body borne well above the ground, their bony and scaly armour, their powerful teeth and their capacious chests, with well-developed ribs, indicate conditions of respiration and general vitality quite comparable with those of the highest modern members of the class Reptilia.

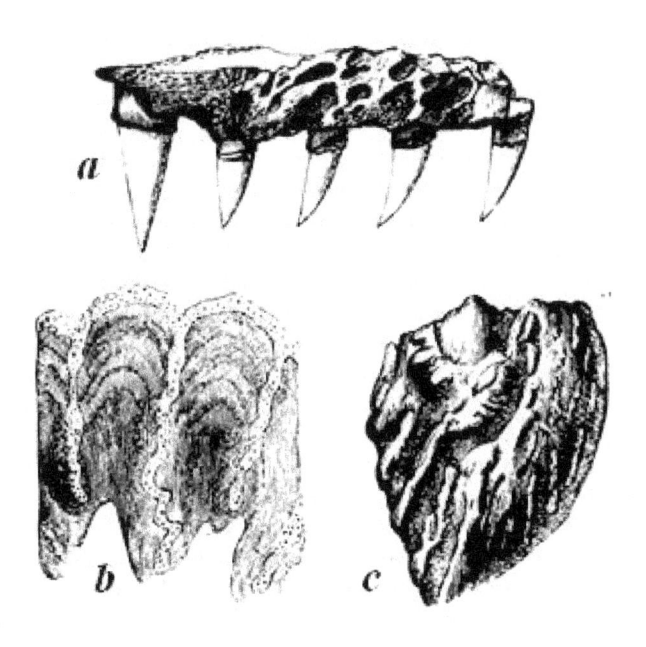

FIG. 137.—*Baphetes planiceps* (Owen).

a, Fragment of maxillary bone showing sculpture, four outer teeth, and one inner tooth. Natural size. *b*, Section of inner tooth. Magnified, *c*, Dermal scale. Natural size.

The third line of progress leads to some slender and beautiful creatures (*Microsauria*), chiefly known to us by remains found in erect trees, and which resembled in form and habits the smaller modern lizards. They have simple teeth, a well-developed brain-case, limbs of some length, and bony and scaly armour, the latter in some cases highly ornate.[60] They were probably the most thoroughly terrestrial, and the most active of the coal batrachians, if indeed they were not strictly intermediate between them and the lizards proper. Fig. 138 shows some fragments of one of these animals; and the animal represented in Fig. 139, recently figured by Fritsch, probably belongs to this group.

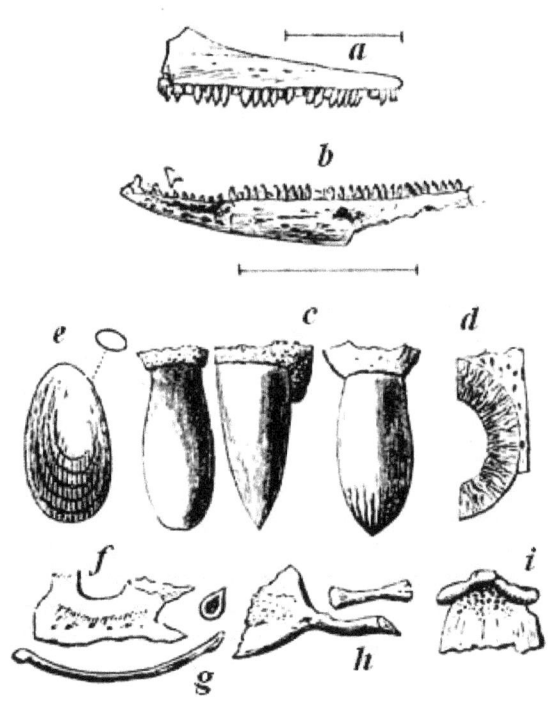

FIG. 138.—A lizard-like Amphibian (*Hylonomus aciedentatus*).

a, Maxillary bone; enlarged.

b, Mandible; enlarged.

c, Teeth; magnified, showing front and side view of ordinary tooth and grooved anterior tooth.

d, Section of tooth; magnified.

e, Scale; natural size and magnified.

f, Pelvic bone (?); natural size.

g, Rib; natural size.

h, Scapular bone (?); natural size.

i, Palate; natural size.

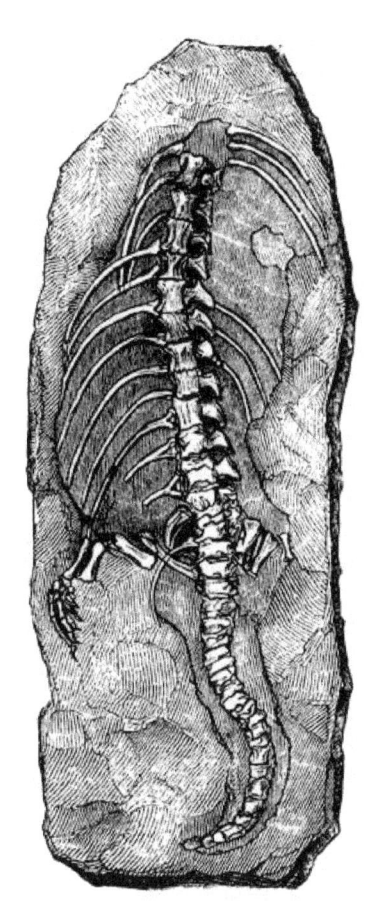

FIG. 139.—*Stelliosaurus longicostatus* (Fritsch). Upper Coal-formation of Bohemia.

The Labyrinthodonts of the Carboniferous continue upward into the Permian, where they meet with the true reptiles; and in the earlier Mesozoic some of the largest and most typical examples are found.[61] But here their reign ceases, and they give place to reptiles of more elevated type, whose history we must consider in the next chapter.

Nothing can be more remarkable than the apparently sudden and simultaneous incoming of the batrachian reptiles in the Coal-formation. As if at a given signal, they came up like the frogs of Egypt everywhere and in all varieties of form. If, as evolutionists suppose, they were developed from fishes, this must have been by some sudden change, occurring at once all over the world, unless indeed some great and unknown gap separates the Devonian from the Carboniferous—a supposition which seems quite contrary to fact—or unless in some region yet unexplored this change was proceeding, and at a particular time its products spread themselves over the world—a supposition equally improbable. In short, the hypothesis of evolution, as applied to these animals, is surrounded with geological improbabilities.

A remarkable picture of the conditions of Palæozoic land life is presented by the occurrence of remains of reptiles, millepedes and land-snails in such erect trees as that represented in Fig. 140. In the now celebrated section of the South Joggins in Nova Scotia, trees of this kind occur at more than sixty different levels; but only in one of these have they as yet been found to be rich in animal remains. Fortunately this bed is so well exposed and so abundant in trees, that I have myself, within a few years, removed from it about twenty of them, the greater number affording remains of land animals.

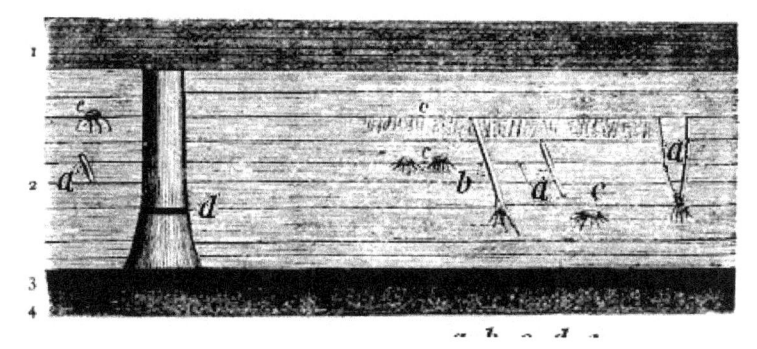

FIG. 140.—Section showing the position of an erect *Sigillaria*, containing remains of land animals.

1. Underclay, with rootlets of Stigmaria, resting on gray shale, with two thin coaly seams.

2. Gray sandstone, with erect trees, *Calamites*, and other stems: 9 feet.

3. Coal, with erect tree on its surface: 6 inches.

4. Underclay with Stigmaria rootlets.

    a, *Calamites*.   *c*, Stigmaria roots.

    b, Stem of plant undetermined.   *d*, Erect trunk, 9 feet high.

The history of one of these trees may be shortly stated thus. It was a *Sigillaria*, perhaps two feet in diameter, and its stem had a dense and imperishable outer bark, a soft cellular inner bark liable to rapid decay, and a slender woody axis not very durable. It grew on the surface of a swamp, now represented by a bed of coal. By inundations and by subsidence, this swamp was exposed to the invasion of muddy and sandy sediment, and this went on accumulating until the stem of the tree was buried to the height of about nine feet, before which time it was no doubt killed. After a time the top decayed and fell, leaving the buried stump imbedded in the sandy soil, which had now become dry, or nearly so. The trunk decayed, its inner bark and axis rotting away and falling in shreds into the bottom of the cylindrical hole, about nine feet deep, once occupied by the stem, and now kept open like a shaft or well by the hard resisting outer bank. The ground around this opening became clothed with ferns and reed-like *Calamites*, partly masking and concealing it. And now millepedes and land snails made the buried trunk a home, or fell into it in their wanderings; and small reptiles sporting around, in pursuit of prey, or themselves pursued, stumbled into the open pitfall, and were unable to extricate themselves, though I have found in some of the layers in these trees trails which show that these imprisoned reptiles had wearily wandered round and round, in the vain search for means of exit, till they died of exhaustion and famine. The bones of these dead reptiles, shells of land-snails and crusts of millepedes, accumulated in these natural coffins, and became mixed with vegetable debris falling into them, and with thin layers of mud washed in by the rains; and this process continued so long that a layer of six inches to a foot in thickness, full of bones, was sometimes produced. At length a new change supervened, the area was again inundated and drifted over with sand, and the hollow trunk was filled to the top and buried under many feet of sediment, never to be re-opened till, after the whole had been hardened into sandstone and elevated to form a part of the modern coast, when the old tree and its forest companions which had shared the same fate with it, are made to yield up their treasures to the geologist. This history is no fancy picture. It represents the results of long and careful study of the beds holding these erect trees, and of the laborious extraction of great numbers of them, and the breaking-up of their contents into thin flakes, to be carefully examined with the lens under a bright light in search of the relics they contained. Fig. 11 in Chap. I. represents the extraction of one of these trees, which happened to be partially exposed by the wasting of the cliff; but many others had to be laboriously mined out of the rock by blasting with gunpowder.

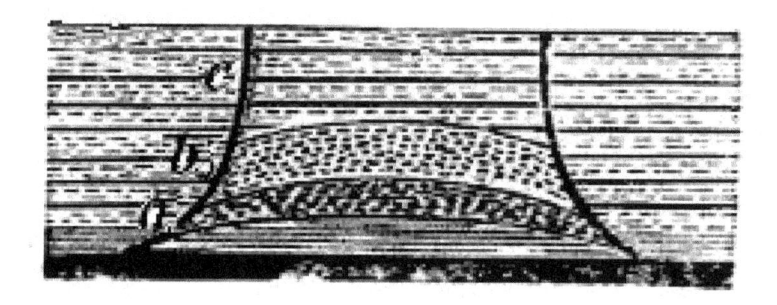

FIG. 140a.—Section of base of erect *Sigillaria*, containing remains of land animals.

a, Mineral charcoal. *b*, Dark-coloured sandstone, with plants, bones, &c. *c*, Gray sandstone, with *Calamites* and *Cordaites*.

It is evident that the combination of circumstances referred to above could not often occur; and it is therefore not wonderful that only in one place and one bed has evidence of it been found, and that even in this some of the trees have been filled up at once by sand and clay, or so crushed by falling in or lateral pressure, that they could receive no animal remains. In one respect this is a striking evidence of the imperfection of the geological record, since, but for what may be called a fortunate accident, many of the most interesting inhabitants of the coal forests might have been altogether unknown to us. On the other hand, it shows how strange and unexpected are the ways in which the relics of the old world have been preserved for our inspection, and that there is probably scarcely any animal or plant that has ever lived of which some fragment does not exist, did we know where to look for it.

It may be well to remark, in closing this chapter, how many new forms of life, air-breathing and otherwise, make their first appearance in the Carboniferous, and have continued to prevail until now. Here we find the first specimens of Amphibians, Spiders, Myriapods, Orthopterous and Coleopterous Insects, and of the Crabs among ten-footed Crustaceans. In the latter group Woodward has recently described the oldest known crab, from the Coal-formation of Belgium.

INHABITANTS OF THE ENGLISH SEAS IN THE AGE OF REPTILES.
Pliosaurus, Ichthyosaurus, Plesiosaurus, Mososaurus, and *Teleosaurus.*

CHAPTER VII.

THE EMPIRE OF THE GREAT REPTILES.

HAD we lived in the Carboniferous period, we might have supposed that the line of the great Labyrinthodont Batrachians would have been continued onward and elevated, perhaps, in the direction of the Mammalia, to which some features of their structure point. But we should have been mistaken in this. The Labyrinthodonts, it is true, extend into the Trias; but there is perhaps a sign of their coming degradation in the appearance in the Permian of the first known mud-eel, a humble Batrachian form allied to the Newts and Water-lizards.[62] Their special peculiarities are dropped in the Mesozoic in favour of those of certain small and feeble lizard-like animals, appearing first in the Carboniferous, and more manifestly in the Permian, and which are the true forerunners, though they can scarcely be the ancestors, of the magnificent reptilian species of the Mesozoic, which have caused this period to be called "the age of reptiles."

The leading reptilian animal from the European Permian has long been the *Proterosaurus*, from the copper slates of Thuringia (Fig. 141), a reptile of lizard-like form, with well-developed limbs, and attaining a length of three or four feet. It resembles more nearly those large modern lizards known as "Monitors," than any other existing form. The fore-limb represented in the figure foreshadows very closely the bones of the human arm and hand. Besides this we find in the Permian certain lizards (*Theriodonts* of Owen) which present the remarkable and advanced peculiarity already predicted by some Carboniferous Microsauria,[63] of having distinct canine teeth, producing a division into incisors, canines, and molars, in the manner of the Carnivorous quadrupeds, which they seem also to have resembled in some other parts of their skeletons. It is not impossible that the footprints in the Permian sandstones of Scotland, which have been referred to tortoises, were those of animals of this type. Cope has recently described from the Permian of Texas a number of reptiles which have the complex dentition of the Theriodonts, and others which simulate that of Herbivorous mammals, by the possession of flat grinding teeth supposed to be adapted to vegetable food.[64] The teeth of all these Permian reptiles were set in sockets, also an advanced peculiarity. Thus already in the Permian, before the final decadence of the Carboniferous flora, and while the Palæozoic invertebrates still lingered in the sea, the age of reptiles dawned, and gave promise of its future greatness by the assumption on the part of reptilian species of structures now limited to the Mammalia.

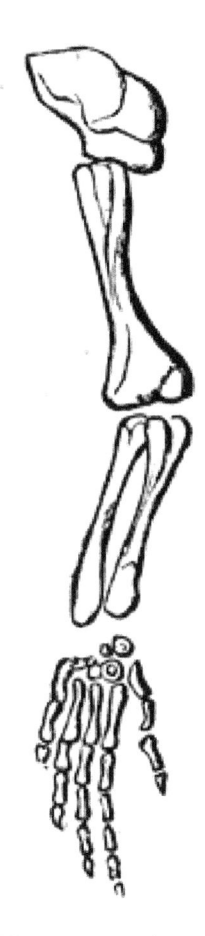

FIG. 141.—Arm of *Proterosaurus Speneri*. Reduced. Permian.

But the great Mesozoic reptiles were not fully enthroned, till the Permian, an unsettled and disturbed age, characterised by great earth movements, had passed away, and until that period of continental elevation, with local deserts and desiccation, and much volcanic action, which we call the Trias, had also passed.

Then in the Jurassic and early Cretaceous the reptiles culminated, and presented features of magnitude and structural complexity unrivalled in later times. At the same time the Labyrinthodonts disappear, or are degraded into the humble stations which the modern Batrachians now occupy.

To understand the reptiles of this age, it will be necessary to notice the subdivisions of their modern representatives. The true reptiles now existing constitute the following orders:—1, the Turtles and Tortoises (*Chelonia*); 2, the Snakes (*Ophidia*); 3, the Lizards (*Lacertilia*); 4, the Crocodiles and Alligators (*Crocodilia*). All of these, except the snakes, are well represented among Mesozoic fossils; but we have in this middle age of the earth's geological history to add to them from five to seven orders now altogether extinct, and these not of low and inferior organisation, but including species far in advance of any now existing both in elevation and magnitude, and constituting the veritable aristocracy of the reptile race. It will best serve our purpose here to consider chiefly these perished orders and their history, and then to notice very shortly those that now survive.

FIG. 142.—Skeleton of *Ichthyosaurus*. Lias. England.

The first of the extinct orders is that of the great Sea-lizards,[65] of which the now familiar *Ichthyosaurus* and *Plesiosaurus* of the English seas, to be seen in all museums and text-books, are the types (Figs. 142, 142*a* and 142*b*). These were marine animals of large size, but not fishes or amphibians. They were true air-breathing reptiles, but with paddles for swimming instead of feet, and some of them with long flattened tails for steering and propulsion. They bore, in short, precisely the same relation to the other members of the class Reptilia which the Whales and Porpoises bear to the ordinary quadrupeds. Some of these animals are believed to have been fifty or sixty feet in length, thus rivalling the Whales, while others were of smaller dimensions, like the Porpoises and Dolphins. Some, like the *Ichthyosaurus* and *Pliosaurus* (Fig. 142*a*), were strongly built and powerful swimmers, and able to destroy the largest fishes, while others, like *Plesiosaurus*, had the body short and compact, the head small, and the neck long and flexible, and probably preyed on small animals near the borders of the waters. Catalogues of British fossils alone include about thirty species of Enaliosaurs, which haunted the coasts of Mesozoic Europe, a wonderful fact, when we consider the absence of these creatures from the modern seas, and the probability that only a fraction of the species are yet known to us.

FIG. 142*a*.—Head of *Pliosaurus*. Jurassic. Much reduced.

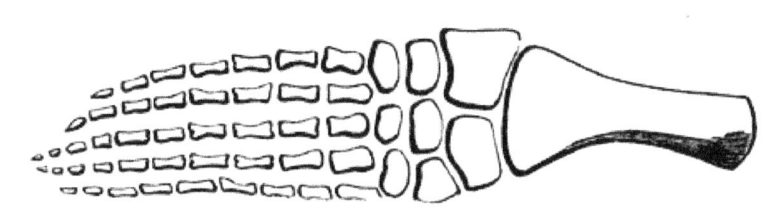

FIG. 142*b*.—Paddle of *Plesiosaurus Oxoniensis*. Jurassic.—After Phillips. One-tenth natural size.

Another remarkable group is that to which Cope has given the name of *Pythonomorpha*, and which he regards as allied to the serpents, or as gigantic sea-serpents provided with swimming paddles, but which Owen considers more nearly connected with the lizards. In either case they constitute a group by themselves, remarkable not only on account of their anatomical affinities with animals so unlike them in general port, but also for their enormously extended length and formidable dentition (Fig. 143). Such animals as the *Mososaurus* of Maestricht and *Clidastes* of Western America may have exceeded in length the largest Ichthyosaurs and the most bulky of living Cetaceans, though their slender forms and numerous vertebræ remind one of the semi-fabulous sea-serpent, rather than of any known animal of our modern age. They were characteristic of the Later Mesozoic, more especially of the Cretaceous period, and must have been formidable enemies to the fishes of their time.

Owen has formed two orders[66] for the reception of some remarkable extinct reptiles of this age, found especially in South Africa and India, but also in Europe and America. The first includes large lizard-like animals having horny jaws like those of turtles, and in some of the species with great defensive tusks (Fig. 144). Their mode of life is not well known, but they may have been peaceable and harmless vegetable feeders. The second has been already referred to, in connection with the Permian, where it first appears, though it is continued in the Trias (Fig. 145). The resemblance of the skulls of these creatures to those of Carnivorous mammals is very striking, and nothing can be more singular than their early appearance and their decadence before the advent of those Tertiary mammals which in more modern times occupy their place.

FIG. 143.—Skeleton of *Clidastes*. A great Mososauroid Sea Reptile of the Cretaceous.—After Cope, much reduced.

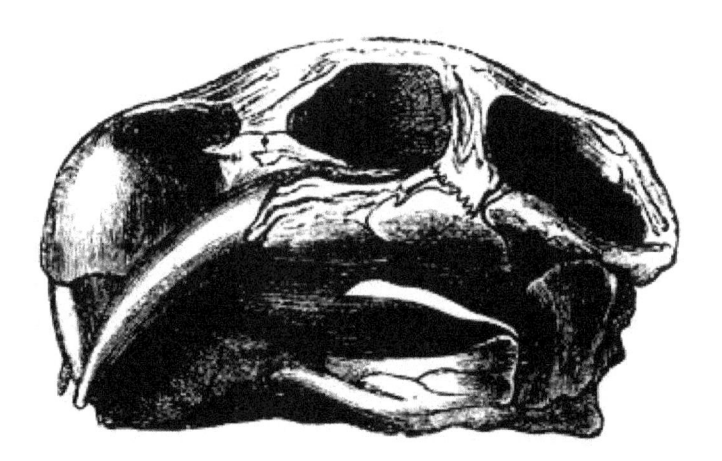

FIG. 144.—An Anomodont Reptile of the Trias (*Dicynodon lacerticeps*, Owen). Reduced.

FIG. 145.—A Theriodont Reptile of the Trias (*Lycosaurus*).—After Owen. Reduced.

FIG. 146.—Skeleton of *Pterodochylus crassirostris*. Jurassic of Solenhofen. Reduced.

FIG. 147.—Restoration of *Rhamphorhyncus Bucklandi*. Jurassic of England.—
After Phillips.

a, One of the teeth. Natural size.

Perhaps the most extraordinary of all the Mesozoic modifications of the reptilian type was that of the flying reptiles, or *Pterodactyls*. These were, in short, lizards modified for flight, somewhat in the same manner with the bats among the mammals. If the bat may be likened to a flying shrew-mouse, a Pterodactyl may in like manner be compared to a flying lizard; but the modification in the latter case is by much the more remarkable, inasmuch as the lizard is a cold-blooded animal, and far less likely to be endowed with the active circulation and muscular power necessary to flight than is the mouse. In point of fact, there can be no doubt that the Pterodactyls must have been provided with some approach to a mammalian or ornithic heart, as they certainly were with great breast-muscles attached to a keel in the breast-bone for working their large membranous wings. These wings were also somewhat original in their construction. They were not furnished with pinions, like those of the bird, but with a membrane like that of the bat, and this, instead of being stretched over four enormously lengthened fingers, as in that quadruped, was supported on a single elongated finger, corresponding, singularly enough, to the little finger, which usually inconspicuous member constituted in some of these strange creatures a limb longer than the whole body (Figs. 146, 147.) The other fingers of the hand were left free for walking or grasping. They are thus believed to have been able to walk as well as to fly, and even in case of need, to swim; while they could probably perch like birds on rocks and trees. Their heads, though very lightly framed, were large and reptilian in aspect, and furnished with sharp teeth, and sometimes probably with a beak as well. Few creatures of the old world are of more hideous and sinister aspect. Yet some of them must have been as light and graceful on the wing as swallows or sea-gulls. There are many species, most of them small, but some of those in the later Mesozoic attained to so great a size that the expanse of their wings must have exceeded twenty feet, making them veritable flying dragons, probably formidable to all the smaller animals of their time. Though these animals were strictly reptiles, they combined in their structures contrivances for aërial locomotion now distributed between the bats and the birds. They had bat-like wings and bird-like chests. Some had horny beaks. All had hollow limb bones, and air cavities to give lightness to the skull. Their brains approach to those of birds, and, as already stated, their respiration and circulation must have been of a high order. These facts are very suggestive, and perhaps in no point is the imagination or the faith of the devout evolutionist more severely tested than in realising the spontaneous assumption of these characters by reptiles, and their subsequent distribution between the very dissimilar types in which they are now continued.

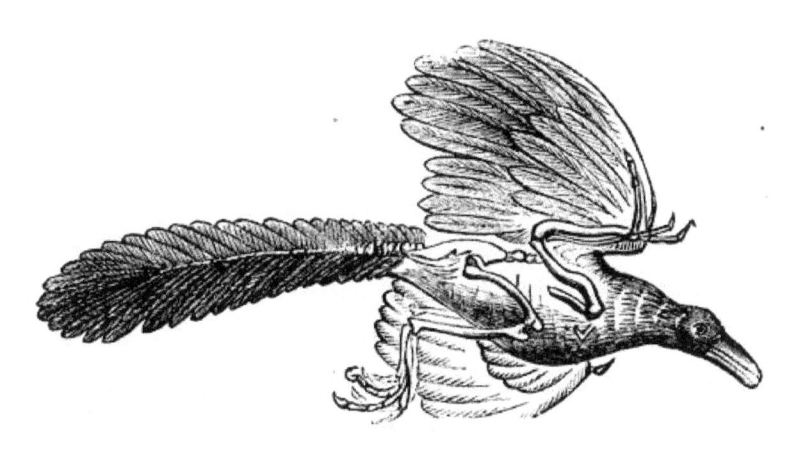

FIG. 148.—A Jurassic Bird (*Archæopteryx macroura*).—After Owen.

FIG. 149.—Jaw of a Cretaceous Toothed Bird (*Ichthyornis dispar*).—After Marsh. Natural size.

The approximation of the winged reptiles to the birds is further increased by the facts that in the Jurassic and Cretaceous periods there were birds having reptilian tails and probably toothed jaws (*Archæopteryx macroura*, Fig. 148). The species just named, while in its limbs, trunk, and feathers a veritable perching bird, resembles a reptile in its head and tail. In the Cretaceous of Western

America, Marsh has recently discovered two distinct types of toothed birds, one having the teeth in regular sockets, the other having them implanted in a groove in the jaw. One of these birds (*Ichthyornis dispar*, Fig. 149) was larger than a pigeon, with powerful wings constructed like those of ordinary birds. It had also the curious and old-fashioned peculiarity of biconcave vertebræ, like those of fishes and some reptiles. Another (*Hesperornis regalis*) stood five or six feet high, and had rudimentary wings like those of the Penguins. These toothed birds extend into the Eocene Tertiary, where the *Odontopteryx* of Owen has been known for some time. In the Eocene, however, this toothed bird is associated with others of ordinary types, allied closely to the Ostriches, the Pelicans, the Ibis, the Woodpeckers, the Hawks, the Owls, the Vultures, and the ordinary perching birds. In the Later Mesozoic, indeed, some reptiles became so bird-like that they nearly approached the earliest birds; but this was a final and futile effort of the reptile to obtain in the air that supremacy which it had long enjoyed in earth and water; and its failure was immediately succeeded in the Eocene by the appearance of a cloud of true birds, representing all the existing orders of the class.

FIG. 150.—Jaw of *Bathygnathus borealis* (Leidy). A Triassic Dinosaur from Prince Edward Island.

a, Cross section of second tooth, natural size. *b*, Fifth tooth, natural size.

We may close our notice of the winged reptiles of the Mesozoic by quoting from Phillips his summary of the characters of *Rhamphorhyncus* (Fig. 147)[67]: "Gifted with ample means of flight, able at least to perch on rocks and scuffle along the shore, perhaps competent to dive, though not so well as a palmiped bird, many fishes must have yielded to the cruel beak and sharp teeth of the *Rhamphorhyncus*. If we ask to which of the many families of birds the analogy of structure and probable way of life would lead us to assimilate Rhamphorhyncus, the answer must point to the swimming races, with long wings, clawed feet, hooked beak, and habits of violence and voracity; and for preference, the shortness of the legs and other circumstances may be held to claim for the Stonesfield fossil a more than fanciful similitude to the groups of Cormorants and other marine divers which constitute an effective part of the picturesque army of robbers of the sea."

FIG. 151.—*Hadrosaurus Foulkii* (Cope). An Herbivorous Dinosaur, 28 feet long.—After Hawkins's restoration.

Lastly, the reptiles, in this age of their imperial sway, culminated in the *Dinosaurians*, animals far above any modern Reptilia in the perfection of their organisation, and many of them of gigantic size. Just as the Pterosaurs filled the place now occupied by the birds, so the Dinosaurs filled that represented by the mammals, or rather they took up a place holding some close relations with both the birds and the mammals. There were thus reptilian animals which on the one hand were the elephants and lions of their time, and on the other bore a grotesque resemblance to creatures so unlike these as the Ostriches, in so far as their anatomical structure was concerned; while it is evident that their whole organisation places them in the highest position possible within the reptilian class. Some of them must have been herbivorous, and probably slow in movement and quiet in nature. Others were carnivorous and of terrible energy, while furnished with the most destructive weapons (Figs. 152, 153). Many had the power of erecting themselves on their hind-feet and walking as bipeds; and to adapt them to this end their hinder limbs were very large and strong, and they had long pillar-like tails, while their fore-feet were comparatively small, and used perhaps mainly for prehension (Figs. 151, 154).

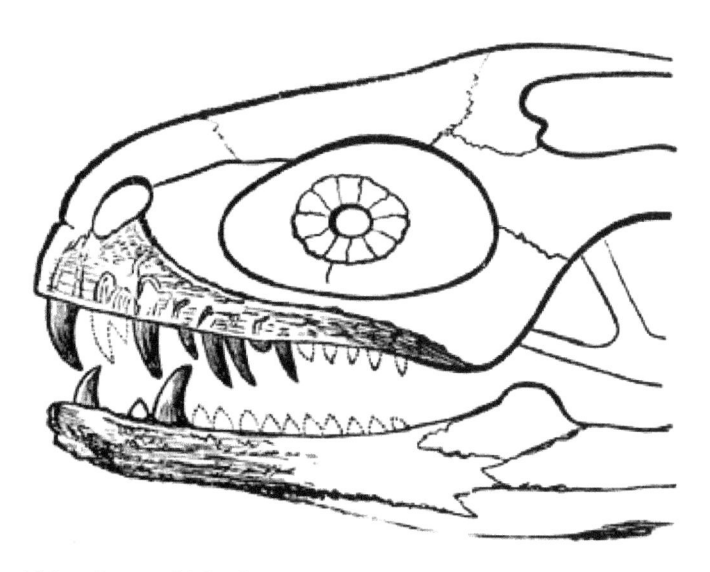

FIG. 152.—Jaws of *Megalosaurus*.—After Phillips. One-tenth natural size.

The size of some of these creatures was stupendous. The *Hadrosaurus* of New Jersey, an Herbivorous species (Fig. 151), when erected on its hind limbs and tail, must have stood more than twenty feet in height. *Megalosaurus* and *Iguanodon*, of the English Jurassic and Wealden, must have been of still more gigantic size. The former was a carnivorous animal, its head (Fig. 152) four or five feet in length, armed with teeth, sabre-shaped, sharp and crenate on the edges (Fig. 153), its hind limbs of enormous power, so that if our imagination does not fail us in the attempt to realise such a wonder, we may even suppose this huge animal, much larger than the largest elephant, springing like a tiger on its prey, a miracle of terrible strength and ferocity, before which no living thing could stand. Its companion, *Iguanodon*, was, on the contrary, a harmless herbivorous creature, using its great strength and stature as a means of obtaining leaves and fruits for food, and perhaps falling a prey to the larger Carnivorous Dinosaurs its contemporaries. A still more bulky animal was the *Ceteosaurus*, so admirably described by Phillips. Its thighbone measures more than five feet in length and a foot in diameter; and it must have stood ten feet high when on all fours, while its length must have reached forty or fifty feet. It seems from the forms of its bones to have been able to walk on land, but probably spent most of its time in the water, where it may be compared to a huge reptilian hippopotamus. Very recently some bones found in rocks, possibly of Wealden age, in Western America, and described by Cope and by Marsh, indicate that even *Ceteosaurus* had not attained to the maximum of Dinosaurian dimensions. These new animals have vertebræ twenty inches in length and from twelve inches to thirteen inches in the diameter of their bodies, while their lateral processes stretched three and a half feet. The shoulder-blade of one species is five feet in length, and its thigh-bone is six feet long. From these measurements Cope concludes that, unlike most other Dinosaurs, it had the fore-feet larger in proportion than the hind-feet, so as to have somewhat the appearance of a large giraffe. The bones of the back have a remarkable cavernous structure, which Cope interprets as indicating air cavities, to give lightness, as in the case of the bones of birds; but Owen suggests that the cavities were filled with cartilage, and that the animals were aquatic in their habits. Evidently in point of size the Dinosaurs had a better claim than even Behemoth to be called the "chief of the ways of God." Some of them, however, were of small size, and probably active and bird-like in their movements. One of these is the animal represented in Fig. 154, a species from the lithographic limestone of Solenhofen.[68]

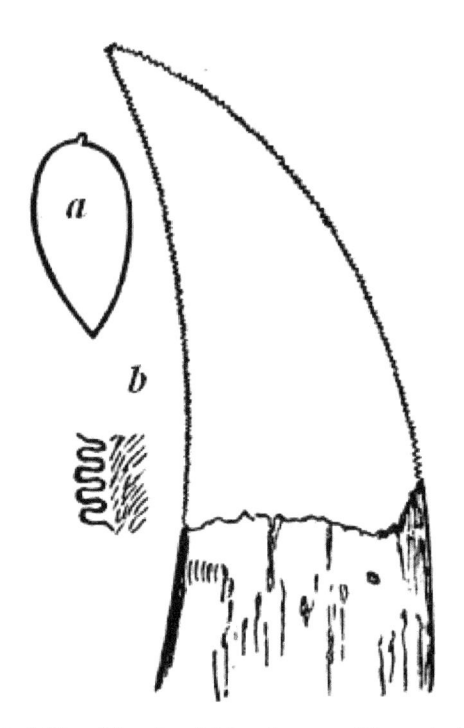

FIG. 153.—Tooth of *Megalosaurus*. Natural size.

a, Cross section. *b*, Crenellation of edges. Enlarged.

Nothing in the life of the Mesozoic has so seized on the imagination of evolutionists as the links of connection between birds and reptiles, which has even been introduced by Huxley into the classification of animals, by his grouping these heretofore very distinct classes in one gigantic and comprehensive class of *Sauropsida*. It is necessary, therefore, to glance at these connections, and if possible to arrive at some conception of their true value. The links which connect the reptiles and the birds are twofold. First, that between the Dinosaurs and the ostrich tribe,[69] and, secondly, that between the Pterodactyls and their allies, and the peculiar Mesozoic birds, such as *Archæopteryx*. The first would serve to account for the few exceptional Struthious birds of the modern world. The second would account for the Passerine and other more ordinary birds; and thus, according to evolution, the now somewhat homogeneous class of birds would have a double, or more probably multiple, origin from several lines of reptilian ancestors. This, no doubt, greatly complicates the links of connection, whether these be supposed to indicate derivation or not.

FIG. 154.—*Compsognathus.* One of the smaller Dinosaurs.—After Wagner.

If we inquire as to the first connection above stated, we may define it briefly in the words of Prof. Phillips, with reference to *Megalosaurus*, which "was not a ground-crawler, like the alligator, but moving with free steps, chiefly, if not solely, on the hind limbs, and claiming a curious analogy, if not some degree of affinity, with the ostrich."[70] But the question arises, Was this resemblance merely that of two oviparous bipeds, or anything more? and when we set off, against the resemblance in haunch bones and hind limbs, the entire dissimilarity in head, in fore limbs, in vertebræ, in tail, and probably in external covering, we are disposed to agree with Huxley in his statement, with respect to the Struthious birds, that their "total amount of approximation to the reptilian type is but small; and the gap between reptiles and birds is but very slightly narrowed by their existence." There is therefore here a great gap, even in the linking together of the types, independently of any question of derivation.

The second line of connection appears at first sight more promising. *Archæopteryx* has a reptilian tail, and claws on the wing; and, as it had toothed jaws, like some of the birds in the Cretaceous, must have altogether made a much nearer approach to a reptile than any modern bird does. The remarkable "fish-bird" (*Ichthyornis*) of Marsh is also very reptilian in some of its characters. But when we compare these reptilian birds with the

Pterodactyls and their allies, a vast gap at once becomes apparent. Disregarding the external clothing, we find the wing in the two groups entirely dissimilar in details of construction, and this dissimilarity extends to the hind limbs as well, so that the Pterodactyls resemble bats rather than birds.

Without committing ourselves to any doctrine of development, we might have rejoiced if our geological discoveries had established a continuous chain, or two continuous chains, of being between the reptiles and the birds; but this end is evidently still far from being attained, though some approximation has undoubtedly been made. To quote again the admission of Huxley: "Birds are no more modified reptiles than reptiles are modified birds, the reptilian and ornithic types being both in reality somewhat different superstructures, raised upon one and the same ground-plan"—that ground-plan being the idea of the air-breathing oviparous vertebrate, and the reptile representing the less specialized and less ornate building. As yet the origin of that idea, and the mode of carrying it out to completion, remain unknown, except to the Architect and Builder, who may reveal them to earnest seekers for truth in His own good time.

As to links of connection with the Mammalia, these are still more obscure. In the Mesozoic the mammals are represented as yet only by a few small species allied to the pouched (Marsupial) and insectivorous quadrupeds of Australia, and these are closely linked with some of the smaller carnivorous Mammalia of the early Tertiary; but neither approach very closely to any known reptilian types. Nor have we yet any connecting links between the great marine reptiles and the Cetaceans and Sirenians which in the Tertiary take their place in the sea.

It is an interesting fact, to come before us in our next chapter, that the great land reptiles of the Mesozoic survived long enough to become contemporary with the introduction and first luxuriance of the modern types of vegetation in the later Cretaceous. It would be natural to suppose that access to these great supplies of better food would have stimulated the increase and development of the herbivorous species, and would have indirectly had the same effect on those that were carnivorous; but the opposite result seems to have followed, and in the next period the reptiles altogether gave place to the mammals, unless, indeed, they were themselves by some mysterious and comparatively rapid process transformed into Mammalia, to suit them to the better conditions of an improved world.

So far as yet known, the reign of reptiles was world-wide in its time; and the imagination is taxed to conceive of a state of things in which the seas swarmed with great reptiles on every coast, when the land was trodden by colossal reptilian bipeds and quadrupeds, in comparison with some of which our elephants are pygmies, and when the air was filled with the grotesque and formidable Pterodactyls. Yet this is no fancy picture. It represents a time which actually existed, when that comparatively low, brutal, and insensate type of existence represented by the modern crocodiles and alligators was supreme in the world. The duration of these creatures was long, and in watching the progress of creation, they would have seemed the permanent inhabitants of the earth. Yet all have perished, and their modern successors, except a few large species existing in the warmer climates, have become subject to the more recently introduced Mammalia.

How did the ancient reptile aristocracy perish? We are ignorant of the details of the catastrophe, but their final disappearance and replacement by the more modern fauna was connected with a great continental subsidence in the Cretaceous age, and with changes of climate and conditions preceding and subsequent to it. Yet the struggle for continued dominion was hard and protracted; and toward its close some of the champions of the reign of reptiles were the greatest and most magnificent examples of the type; as if they had risen in their might to defy approaching ruin. Thus some of the most stupendous forms appear in the later Cretaceous, after the great subsidence had made progress and almost attained its consummation. Like the antediluvian giants, they were undismayed even when the land began to sink beneath their feet; and for them there was no ark of deliverance.

LOWER CRETACEOUS LEAVES. REDUCED IN SIZE.—After Lesquereux.

a, *Aralia Saporteana*. *b*, *Sassafras araliopsis*. *c*, *Quercus primordialis*. *d*, *Fagus polyclada*. *e*, *Salix proteæfolia*. *f*, *Laurus proteæfolia*.

CHAPTER VIII.

THE FIRST FORESTS OF MODERN TYPE.

FOR a long time it was believed by geologists that a great and mysterious gap separated the Upper Cretaceous from the oldest Tertiary formations; and in Western Europe, in so far as physical conditions and animal life are concerned, the severance seemed nearly complete. Oceanic deposits, like the Upper Chalk, are succeeded by beds of littoral and estuarine characters. The last and some of the greatest of the Mesozoic Saurians have their burial-places in the Upper Cretaceous, and appear no more on earth. The wonderful shell-fishes of the Ammonite group, and the cuttle-fishes of the Belemnite type, share the same fate. With the earliest deposits of the Eocene Tertiary came in multitudes of large Mammalia heretofore unknown, and the Cetaceans appear in the sea instead of the great marine lizards; while shells, corals, and crustaceans of modern types swarm in the waters. Thus it is true that a great and apparently somewhat abrupt change takes place at the close of the Cretaceous, and terminates for ever the reptilian age. Even in regions like Western America, where physically the later Cretaceous shades gradually into the earlier Tertiary, so that there have been doubts as to the limits of these several periods, the same great change in animal life occurs.

But a link of connection has at length been found in the history of the vegetable kingdom. The modern flora came in with its full force in the later Cretaceous, before the end of the reptilian age, and continued onward to the present time. Thus the plant takes precedence of the animal, and the preparation was made for the mammalian life of the Eocene by the introduction of the modern flora in the Cretaceous period. In like manner it is possible that the great graphite deposits of the Laurentian indicate a vegetation which preceded the swarming marine life of the Cambrian; and it is not improbable that the Palæozoic land flora existed long before the first land animals. Thus the plant, as in the old Mosaic record, ever appears on the day before the animal, in each stage of the development of the world.

In Chapter IV. we traced the history of the old and rich vegetation of the Coal period. But this vegetation consisted principally of cryptogams and those lowest phænogams, of the pine and cycad groups, which have naked seeds. In the modern flora we may arrange the several groups of plants, somewhat naturally, as follows:—

Series I., CRYPTOGAMS:—

Class 1, *Thallophytes*, sea-weeds, lichens, fungi.

" 2, *Anophytes*, mosses, &c.

" 3, *Acrogens*, ferns, lycopods, horsetails.

Series II., PHÆNOGAMS:—

Class 4, *Gymnosperms*, pines, cycads, &c.

" 5, *Endogens*, palms, grasses, &c.

" 6, *Exogens*, oaks, maples, &c.

With reference to the history of these groups the record stands as follows:— In the Palæozoic age classes 3 and 4 culminated, and constituted the great mass of the arboreal vegetation. On entering the Mesozoic, No. 3 becomes somewhat diminished, but No. 4 continues with unabated prevalence, so that the Mesozoic has sometimes been characterized as emphatically the age of Gymnosperms. With these appear some Endogens, allied to the modern Yuccas and Screw pines and Arums. But in the lower Mesozoic rocks we have no representatives of the broad-leaved Exogens (Angiosperms), which constitute the great mass of ordinary forest vegetation; and it is only in the Cretaceous that we find them appearing in force, and that the monotonous vegetation of the older style was replaced by the more beautiful and varied forms of our modern woods.

In Europe, in the lower part of the Upper Cretaceous of Bohemia (*Cenomanian*), have been found some leaves which indicate the beginning of this change. These have been referred to Cæsalpinias or Brasilettos, pod-bearing trees of India and tropical America, Aralias or Ginsengs, Magnolias, Laurels, an Ivy, and a peculiar and uncertain genus (*Credneria*). With these are noble palms, both of the types with pinnate and palmate leaves, and trees allied to the Giant Sequoias of California, and to the Araucarian pines of the southern hemisphere. (See Frontispiece to this Chapter.) These ancient Cretaceous forests of Eastern Europe are compared by Saporta with those which now live in the warmer portions of China or in South America—truly a marvellous change from the sombre and uniform vegetation by which they seem to have been immediately preceded. A still further development of modern vegetation takes place in the next or highest member of the Cretaceous, the Maestricht beds (*Senonian*), where we find a crowd of modern types. On this great change Count Saporta remarks with truth that there seem to have been periods of pause and of activity in the introduction of plants. The Jurassic period was one of inactivity; and a new and vigorous evolution, as he regards it, is introduced in the middle of the Cretaceous.

This new and grand elevation of the vegetable kingdom in the Cretaceous age was not local merely. In Moravia, in the Hartz, in Belgium and France, even in Greenland, the same great renewing of the face of the earth was in progress. In America it was proceeding on a grand scale, and seems to have set in earlier than in Europe.[71] In the Dakota group of the West, one of the lower members of the Cretaceous, and covering a vast area, a rich angiospermous flora has been discovered by Hayden, and described by Lesquereux and Newberry, and beds of coal have been formed from its remains. In Vancouver's Island in British Columbia, Cretaceous coal measures occur, comparable in value and in the excellence of the fuel they afford with those of the true coal formation. Some of the beds of coal are eight feet in thickness, and the shales associated with them abound in leaves of exogenous trees generally similar to those still living in America. In these beds are also found mineralized trunks, which present under the microscope the familiar structures of our oaks, birches, and other modern trees. Thus all over the northern hemisphere the elevation of the land out of the waters of the great Cretaceous subsidence was signalized by a development of noble and exuberant forest vegetation, of the types still extant. The following list of families found in the Cretaceous, after Saporta, will show the botanist how fully our modern Exogens are represented:—

APETALÆ.	GAMOPETALÆ.	POLYPETALÆ.
Myricaceæ.	*Apocynaceæ.*	*Araliaceæ.*
Cupuliferæ.	*Ericaceæ.*	*Hamameliaceæ.*
Betulaceæ.	*Ebenaceæ.*	*Helleborineæ.*
Salicaceæ.	*Myrsineæ.*	*Magnoliaceæ.*
Moreæ.		*Tiliaceæ.*
Proteaceæ.		*Celastraceæ.*
Lauraceæ.		*Anacardiaceæ.*
		Myrtaceæ.

Of the plants in this list, some, like the oaks, birches, willows, and heaths, are common and familiar members of the flora of the northern hemisphere to-day, and even of the European flora. Some, like the Magnolias, Myricas, and witch-hazels, are characteristically American, and a few, like the Proteaceæ, are now confined to the southern hemisphere. Some of these families have dwindled since the Cretaceous time, so as to be represented by very few species, or at least have not advanced, while others have multiplied and prospered; and on the whole the flora of the northern hemisphere seems to have been as rich in this early beginning of our modern forests as it is at the present day. Lesquereux's results, with reference to the American flora of the Dakota group, are very similar, and present some surprising features of resemblance to modern American forests, though he remarks that these Cretaceous trees are generally characterized by the even or unserrated edges of their leaves; and the same remark seems to apply to the oldest Cretaceous leaves of Europe.

A very singular feature of the Cretaceous flora is the number of species of some genera now represented by few or even a single species; and this is the more remarkable when we consider how few species, comparatively, of the older flora, are known to us. For example, Lesquereux, though aware of the great variability of the modern Sassafras of America, recognizes eight species of this genus in the Dakota Cretaceous, one of which seems to be that still living in America, so that it has continued unchanged, while the others have perished (Fig. 155). Thus this genus culminates at once in the Cretaceous, but continues still in one of its species. Again, the tulip-tree, *Liriodendron*, one of the most beautiful, unique, and invariable of American trees, is represented by one sole species in the present world. There seem to be no less than four in the Dakota beds, besides others in the Cretaceous of New Jersey, and one species is found in the Tertiary of Greenland as well as in that of Europe (Fig. 156). There are probably four or five species of plane-tree (*Platanus*) now extant, of which but one occurs in America, unless *P. Mexicana*, the Mexican plane-tree, is a good species as distinct from the ordinary, more northern, form. There are seven species, according to Lesquereux, in the Cretaceous of Dakota alone. This sort of evolution backward, or from many species to few, would probably be greatly increased, had we fuller knowledge of the Cretaceous flora, as there are several genera already represented by as many species as they can boast in modern times. We have already seen that this abrupt and sudden culmination of genera and families, and their subsequent decadence, is no rare thing in geology, and it connects itself with that idea of periods of creative activity which we have already had occasion to notice.

FIG. 155.—*Sassafras cretaceum* (Newberry).

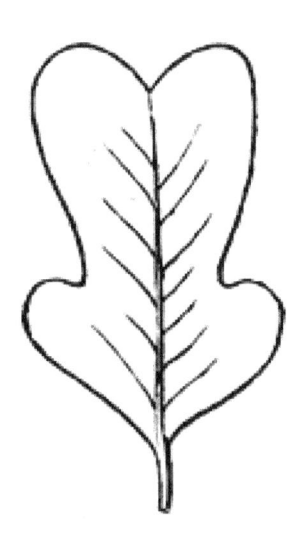

FIG. 156.—*Liriodendron primævum* (Newberry). A Cretaceous Tulip-tree.

FIG. 157.—*Onoclea sensibilis.* Eocene.—After Newberry.

FIG. 158.—*Davallia tenuifolia.* Eocene.—After Dawson. Natural size and enlarged.

I have dwelt principally on the phænogamous plants of the Cretaceous, as presenting the most noteworthy and new features of the time; but we must not forget that though cryptogams were deposed from the high position they held in the Palæozoic, they still existed; and there are more especially many interesting species of ferns and equisetums in the Cretaceous and Eocene rocks. These are, however, of modern types; and it is remarkable that some of them appear to have continued without even specific change from the later Cretaceous up to the present time. A striking illustration of this is afforded by two ferns discovered side by side in the oldest Eocene beds[72] of the plains west of Red River, and described in Dr. G. M. Dawson's report on the 49th parallel. One of these is the well-known and very common *Onoclea sensibilis* (Fig. 157), or sensitive fern of Eastern America.[73] This species came into existence at latest at the close of the Cretaceous, and has apparently been continued in America up to the present time. In Europe, where it does not now live, it occurs as a fossil in Eocene beds in the Isle of Mull. The other is *Davallia tenuifolia* (Fig. 158), a delicate little plant belonging to a genus not now represented in America, and to a species at present found only in Asia. Yet this species also lived in America in early Eocene times, but has since been banished, though its former companion, the *Onoclea*, still holds its ground. Such cases of specific persistence along with great changes of habitat are very instructive as to the permanence of species.

Count Saporta, whose just remarks on the marvellously sudden incoming of the Cretaceous flora we have already referred to, also notices the fact that the families and genera represented in this flora are a most miscellaneous and unconnected assemblage, showing either the simultaneous appearance of many dissimilar types, or requiring us to believe in the existence of these and of intermediate forms for a very long period before that in which they are first found. This may, however, be placed in connection with the appearance of an exogenous tree (*Syringoxylon*) in the Devonian, referred to in a previous chapter. It would be a strange and now little suspected case of imperfection of the record, if it should be found that trees of this type were lurking in exceptional corners through all the vast periods between the Devonian and the Cretaceous, to burst forth in unwonted variety and luxuriance in the latter period.

The new Cretaceous flora appears first in beds which had been recently elevated from the ocean of the great Cretaceous subsidence; and when it first flourished, in temperate regions at least, the continents were of small dimensions, and broken up into groups of islands. Farther, America would seem to have had precedence of the Eastern Continent, and the Arctic of the Temperate regions. Thus on the elevation of the later Cretaceous land, plants previously established in the far north spread themselves southward, over newly-raised lands, radiating from the polar regions into Europe, Asia, and

America. This seems the only way of accounting for the similarity of the plants in these distant countries. The new flora of the Upper Cretaceous in its journey southward met with a climate probably warmer than the present, yet not so warm as to prevent trees similar to those now living in the same latitudes from flourishing.

Let us now trace this flora through the succeeding ages, in which I shall follow pretty closely some general statements made by Count De Saporta in memoirs recently published.

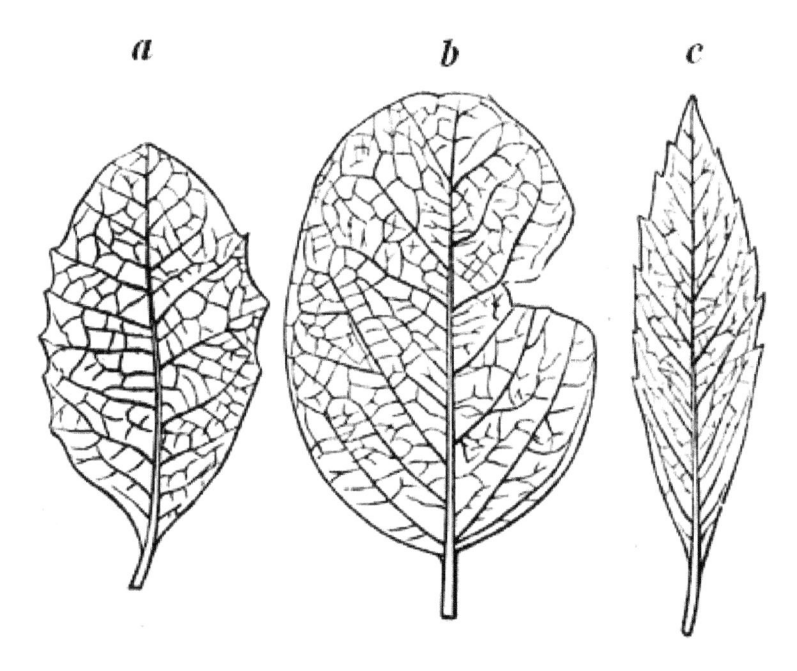

FIG. 159.—Eocene Leaves. From Aix.

a, Quercus antecedens (Saporta). *b, Diospyros pyrifolia* (Saporta). *c, Myrica Mathesonii* (Saporta).

At the beginning of the Eocene we find a humid and warm climate in Europe, with great forests of oaks, chestnuts, laurels, giant pines, and other genera, some of them still European, others Asiatic or American, and many of them survivors of the Cretaceous (Figs. 159 to 162); and at the same period similar forests overspread those great plains of North America which were rising from out the Cretaceous sea, and there vast swampy beds were formed of vegetable *débris*, giving origin to beds of brown coal, some of them eighteen feet in thickness. Then came in Europe and Asia that great subsidence under the sea, during which the Nummuline limestones were deposited, and when the old continent was resolved again into an archipelago of islands, perhaps closely connected with more southern lands. This led to a great increase of southern forms of plants, which does not seem to have occurred to the same extent in America, where the flora is more continuous, though showing a warmer climate in the older than in the newer Eocene. At this period Palms, Screw pines, Proteaceous shrubs, Myrtles, Acacias, and other plants of the character of those of more southern climates were dominant in Europe (Fig. 163). The well-known beds of Bournemouth, in the south of England,[74] contain a rich flora of the Eocene age, perhaps of its middle period, and reminding us of the forests of sub-tropical India or Australia.

FIG. 160.—An Ancient Clover (*Trifolium palæogæum*, Saporta). Eocene. Aix.

FIG. 161.—An Eocene Maple (*Acer sextianus*, Saporta). Aix.

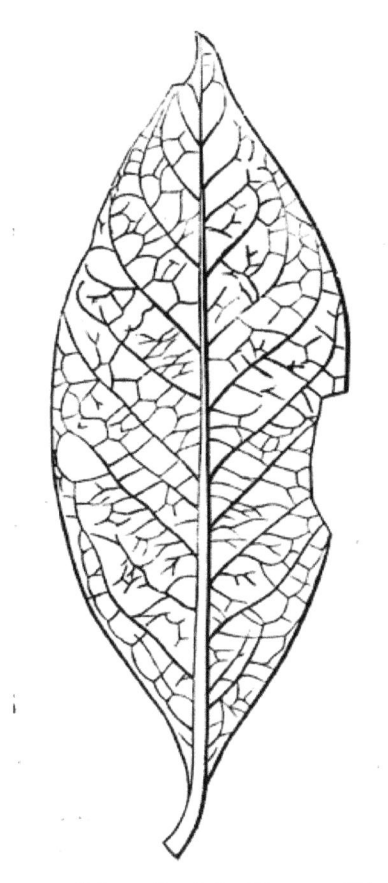

FIG. 162.—A European Magnoli of the Eocene (*M. dianæ*, Saporta). Aix.

FIG. 163.—Flower and Leaf of *Bombax sepultiflorum*. Eocene of Aix.—After Saporta.

A European representative of the Silk-cotton-tree of the East Indies and Tropical America.

Gradual elevation of the land favoured for a time the extension of these plants, and the warmth of the climate allowed them to extend even into Arctic latitudes. But in the middle of the Eocene another subsidence occurred, which exterminated much of the Eocene flora, and was perhaps accompanied with a reduction of temperature, in which the more northern lands became covered with great forests of trees allied to the Pines. In England a remarkable deposit of this age is that of Bovey Tracey, in Devonshire, where beds of clay and brown coal have afforded a rich flora of American and southern types. The *Sequoia* shown in Fig. 164 abounds at this place, and is a near relation to the celebrated "big trees" of California; the *Cinnamomum* in Fig. 165 is a type equally foreign from modern England. It is a curious feature of the Bovey deposit that immediately above these Eocene beds, holding a rich flora of warm temperate character, are glacial clays with leaves of Arctic willows and of the dwarf birch, indicating a climate much more severe than that of the British Islands at present.[75]

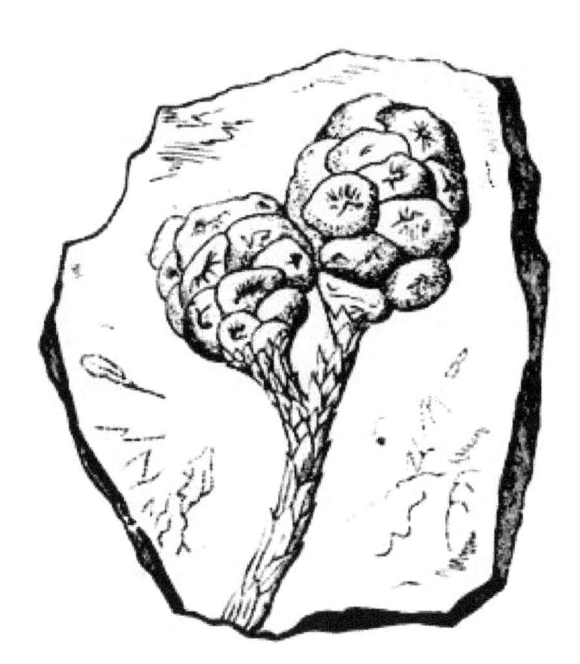

FIG. 164.—Branch and Fruit of *Sequoia Couttsiæ* (Heer). Eocene. England.

In the Miocene period the land again rose, and the northern flora spread itself southward equally over Europe, Asia, and America, so that the Miocene flora of all these regions is very similar; and this Miocene flora continues substantially to this day in Eastern America and Eastern Asia, except that it has been greatly reduced in number of species by the intervention of the cold glacial period; but in Europe and Western America it has been largely replaced by other apparently more modern species.

FIG. 165.—*Cinnamomum Scheuchzeri* (Heer). Eocene. England.

A striking result of recent discoveries is the fact that in Cretaceous and Eocene times a very warm climate prevailed in the extreme Arctic regions, and trees of temperate latitudes grew there freely. In the recent Arctic expedition, Captain Fielden found in latitude 81° 40', within 600 miles of the Pole, a bed of lignite, from twenty-five to thirty feet in thickness, associated with remains of plants such as now grow only in temperate latitudes.

"From the character of the plant-remains, Dr. Heer infers that the lignite of this locality represents an ancient peat-moss, which must have been of wide extent, with reeds, sedges, birches, poplar, and certain conifers growing on its banks; while the higher and drier ground in the neighbourhood probably supported a growth of pines and firs, with elms and hazel-bushes. The remains of water-lilies suggest the existence of a fresh-water lake in the old peat-moss, which must have remained unfrozen during a great part of the year."

It is to be observed with reference to the age of these beds, that as the Later Cretaceous and Eocene flora of Europe and America migrated from the north, the plants found in the beds of that age in the temperate latitudes may really be somewhat older in the Arctic regions, a fact which produces some uncertainty as to their actual age.

The warmth required for the growth of luxuriant forests near the Pole might be secured by a different distribution of land and water, and of the oceanic currents, but the requirements of plants as to light seem more difficult to meet, and it has been doubted whether species similar to those which are accustomed in modern times to regular alternations of day and night could submit to the long Arctic winter darkness. It is known, however, that in conservatories in Northern Russia plants supplied with heat and moisture can endure in winter great deprivation of light, and at Disco, in Greenland, roses and fuchsias flourish as house plants.[76] These facts show that if there were sufficient light and heat in summer, a great number of the plants of temperate latitudes could endure extreme cold and much deprivation of light in winter.

It may be well here to inform the reader that some confusion as to the succession of the Cretaceous and Tertiary floras in America has arisen from the fact that the plants which are evidently Eocene in Greenland and America have been until lately incorrectly regarded as Miocene in Europe. In the

Western States, the Dakota group of Lesquereux is overlain by 2000 feet of Cretaceous beds, containing the marine shells characteristic of that age, but no plants. But in Vancouver's Island these same Upper Cretaceous beds contain an abundant flora, which some botanists have called Tertiary for the reason already mentioned. Above the 2000 feet of marine beds overlying the Dakota group is the Lower Lignite group of Lesquereux, holding many fossil plants, including Palms and other evidences of a warmer climate than that of the Cretaceous, and which constitute a Lower Eocene flora corresponding in some respects to that of Europe. This is succeeded by an Upper Lignite group, also Eocene, but representing a more temperate climate, and therefore resembling more nearly the Cretaceous flora. This is nearly identical with the so-called Miocene of Greenland, Alaska, and Mackenzie River, which the facts collected by the Canadian geologists have shown to be really Eocene.[77] But the Canadian reports containing these facts are comparatively little known in Europe, hence incorrect ideas as to the succession of these floras have been handed from one writer to another.

To those who adopt extreme views as to the refrigeration of the northern hemisphere in so-called glacial times, there is great difficulty in accounting for the continued existence of the early Tertiary flora; but if we adopt moderate views as to this, and demand merely a great subsidence, with much reduction of mean temperature, we may suppose that the plants previously existing were preserved on insular spots, whence they were ready to recolonize the land on its emergence from the sea. It seems certain, however, that our continents never regained, after the Glacial period, the exuberance of plant life which they presented in the Miocene and earlier Pliocene; and we shall find that this statement applies to the world of animals as well as to that of plants. This reduction was more extreme in Europe than in Eastern Asia and Eastern

America, and the fact is thus accounted for in a recent lecture by Prof. Asa Gray:—

"I conceive that three things have conspired to this loss. First, Europe, hardly extending south of latitude 40°, is all within the limits generally assigned to severe glacial action. Second, its mountains trend east and west, from the Pyrenees to the Carpathians and the Caucasus beyond, near its southern border; and they had glaciers of their own, which must have begun their operations, and poured down the northward flanks, while the plains were still covered with forest, on the retreat from the great ice-wave coming from the north. Attacked both on front and rear, much of the forest must have perished then and there. Third, across the line of retreat of those which may have flanked the mountain-ranges, or were stationed south of them, stretched the Mediterranean, an impassable barrier. Some hardy trees may have eked out their existence on the northern shore of the Mediterranean and the Atlantic coast. But we doubt not, *Taxodium* and Sequoias, Magnolias and Liquidambars, and even Hickories and the like, were among the missing. Escape by the east, and rehabilitation from that quarter until a very late period, were apparently prevented by the prolongation of the Mediterranean to the Caspian, and thence to the Siberian ocean. If we accept the supposition of Nordenskiöld, that, anterior to the Glacial period, Europe was 'bounded on the south by an ocean extending from the Atlantic over the present deserts of Sahara and Central Asia to the Pacific,' all chance of these American types having escaped from or re-entered Europe from the south and east is excluded. Europe may thus be conceived to have been for a time somewhat in the condition in which Greenland is now, and indeed to have been connected with Greenland in this or in earlier times.[78] Such a junction, cutting off access of the Gulf Stream to the Polar Sea, would, as some think, other things remaining as they are, almost of itself give glaciation to Europe. Greenland may be referred to, by way of comparison, as a country which, having undergone extreme glaciation, bears the marks of it in the extreme poverty of its flora, and in the absence of the plants to which its southern portion, extending six degrees below the Arctic Circle, might be entitled. It ought to have trees, and might support them. But since destruction by glaciation no way has been opened for their return. Europe fared much better, but suffered in its degree in a similar way.

"Turning for a moment to the American continent for a contrast, we find the land unbroken and open down to the tropic, and the mountains running north and south. The trees, when touched on the north by the on-coming refrigeration, had only to move their southern border southward, along an open way, as far as the exigency required; and there was no impediment to their due return. Then the more southern latitude of the United States gave great advantage over Europe. On the Atlantic border, proper glaciation was felt only in the northern part, down to about latitude 40°. In the interior of the country, owing doubtless to greater dryness and summer heat, the limit receded greatly northward in the Mississippi Valley, and gave only local glaciers to the Rocky Mountains; and no volcanic outbreaks or violent changes of any kind have here occurred since the types of our present vegetation came to the land. So our lines have been cast in pleasant places, and the goodly heritage of forest-trees is one of the consequences.

"The still greater richness of North-east Asia in arboreal vegetation may find explanation in the prevalence of particularly favourable conditions, both ante-glacial and recent. The trees of the Miocene circumpolar forest appear to have found there a secure home; and the Japanese islands, to which most of these trees belong, must be remarkably adapted to them. The situation of these islands—analogous to that of Great Britain, but with the advantage of lower latitude and greater sunshinetheir ample extent north and south, their diversified configuration, their proximity to the great Pacific gulf-stream, by which a vast body of warm water sweeps along their accentuated shores, and the comparatively equable diffusion of rain throughout the year, all probably conspire to the preservation and development of an originally ample inheritance."

The comparative paucity in species of the west coast of America, though the Sequoias and some other forms which have perished elsewhere are retained there, is admitted to be exceptional, and not easily explained, except by the supposition of peculiar local conditions affecting the comparatively narrow strip of land between the Rocky Mountains and coast ranges, and the Pacific.

To such widely-distributed and varied and complex phenomena as those which have been discussed in the present chapter, it is impossible to do justice in the space at our command. Details in relation to them will be found in the publications of Heer, of Saporta, and of Lesquereux, and are well worthy of study by botanists, to whom alone they can be made fully intelligible. In general, with reference to now prevalent theories of derivation, they present two very dissimilar aspects. No difficulty can be greater to the evolutionist than to account for the simultaneous appearance of so many modern generic forms in the Cretaceous; and the fact of many of the genera presenting more and more species the farther we trace them back is a strange anomaly of evolution. On the other hand, the number of species continuing unchanged from the Eocene to the Modern, the others only slightly modified, and the representative species occurring in the floras of the old and new continents, appear to many to give great support to the doctrine of gradual transformation of species. Farther facts and farther comprehension of the difference between species and races will be necessary to the settlement of these questions. In the meantime it would appear that the Jurassic flora rapidly gave place, at a particular point of geological time, to that of the modern world, and this not merely in one locality, but over the whole northern hemisphere; and there are apparently similar facts in the southern hemisphere as well. It farther appears that each genus was at first represented by many species, and that as time went on these were gradually reduced to a few best suited to survive; and that the changes of climate and level which occurred distributed these over different parts of the continents in a way at first sight very anomalous, but which Prof. Gray somewhat quaintly represents as follows:—

"It is as if Nature, when she had enough species of a genus to go round the four floral regions (Europe, East Asia, West America, and East America), dealt them fairly one at least to each quarter of our zone; but when she had only two of some peculiar kind, gave one to us, and the other to Japan, Mantchuria, or the Himalayas; and when she had only one, divided it between the two partners on the opposite sides of the table."

Lastly, it seems very probable that many so-called species are nothing more than varietal forms, which may very well be modified descendants of Miocene or Eocene plants now figuring in our lists under different names.

SIVATHERIUM GIGANTEUM.

A Great Ruminant of the Miocene of India.

Copied by special permission of James Murie, M.D., F.G.S., &c.

CHAPTER IX.

THE REIGN OF MAMMALS.

THE incoming of that highest order of animals in which man himself, in so far as his physical nature is concerned, takes his place, presents some features which, though not unparalleled in the history of other forms of life, are still very striking. The modern Mammalia are somewhat sharply divided into three very unequal groups. First, those which present in their full perfection the property of producing fully developed young, which is one of the distinctive characters of the class. These are the Placental Mammals. Secondly, those in which the young are produced in a very imperfect condition, and are usually nourished for a time in a marsupium or pouch. These are hence called Marsupials. They are for the most part confined to Australasia, though a few occur in America; and are decidedly inferior in rank to the ordinary mammals. Thirdly, those in which there is a bird-like bill, and also certain bird-like or reptilian peculiarities of skeleton and of the alimentary canal. These are the Monotremes, represented by a very few species in Australia and New Guinea.

In geological history, so far as the facts are at present known, the second group, that of the Marsupials, antedated the others by a vast lapse of time. The Marsupials appear in the Trias, near the beginning of the Mesozoic period. The Placentals are not found until we reach the beginning of the Tertiary. The Monotremes would seem to be a comparatively modern degraded type. Thus the Marsupials existed throughout the reptilian age, and this in those countries of the northern hemisphere in which they are not now found. The Mesozoic Marsupials were, it is true, of small size, but there were probably numerous species, and though unable to cope with the great reptiles that swarmed by the shores and on the plains, they may have found abundant scope in the upland and interior regions of the continents.

The Upper Trias of Germany has afforded to Professor Pleininger two teeth of a small mammal, to which the name of *Microlestes antiquus* has been given, under the impression that it was carnivorous, though it now seems more likely that it was a vegetable feeder. In rocks of nearly the same age in America, Emmons found a jaw-bone of another species (*Dromatherium sylvestre*), which has been supposed to be a near ally of the existing *Myrmecobius fasciatus* of Australia (Figs. 166, 167). In the Stonesfield slate, a member of the English Jurassic, several other species have been found (Fig. 168), and a still larger number in the freshwater beds of the Upper Purbeck. Marsh has obtained many others from the Jurassic of America. None appear to have yet been found in the Cretaceous, but they reappear in the Eocene Tertiary, and continue to the modern time. Their absence in the Cretaceous is probably a mere accident, and they present an illustration of a very permanent type little changed since its first introduction. Lyell enumerates in all thirty-three species from the Mesozoic, all of them of small size, and all more or less nearly related to existing Australian Marsupials, though differing much among themselves, and including both carnivorous and herbivorous forms (Fig. 169). Marsh has recently suggested a somewhat new interpretation of these interesting mammalian remains.[79] He considers them divisible into two groups, one allied to the modern Insectivora (Moles, Shrews, Hedgehogs, &c.), but of generalized forms. For these he constitutes a new order (*Pantotheria*, Marsh). The other group is less numerous and is Marsupial (*Allotheria*, Marsh). The jaws in Figs. 166 and 168 belong to the former group, that in Fig. 169 to the latter. We should thus have both placental and Marsupial mammals in the Mesozoic. Marsh remarks that the descent of these different types from a common ancestry would require us to trace mammals back into the Palæozoic, that is, on the doctrine of gradual evolution.

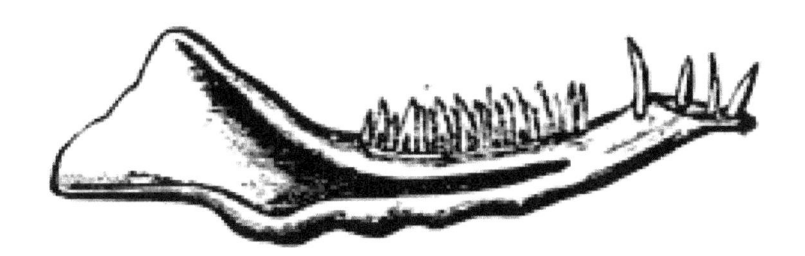

FIG. 166.—Jaw of *Dromatherium sylvestre* (Emmons). From the Trias of North Carolina.

FIG. 167.—*Myrmecobius fasciatus*. A modern Australian marsupial, allied to Mesozoic species.

So soon as the palæontologist passes from the Upper Cretaceous to the Eocene, he finds himself in the domain of the placental mammals, which appear in numerous and large species, and this, not merely in one region, but in every part of the world in which these deposits are known to exist.

FIG. 168.—Jaw, and enlarged molar of *Phascolotherium Bucklandi*. Stonesfield slate. England.—After Phillips.

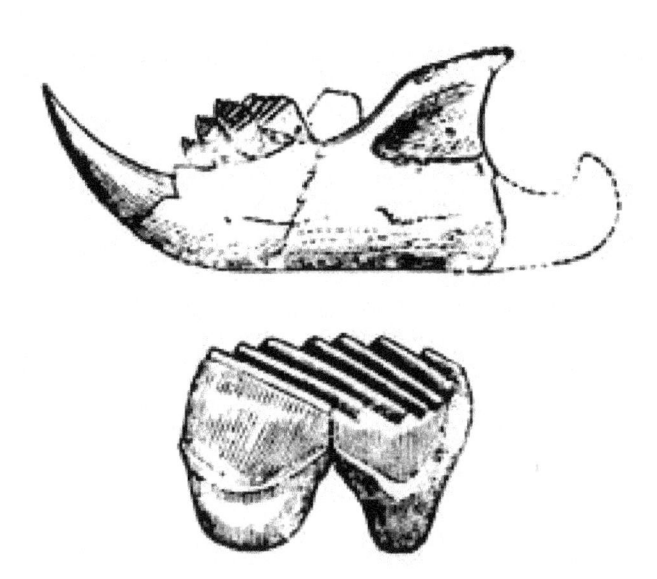

FIG. 169.—*Plagiaulax Becklesii*. Jaw, and pre-molar enlarged, showing flat surface, with ridges.—Purbeck.

Indeed, the recent discoveries in America and in the east of Europe have almost thrown into the shade those researches of Cuvier in the Paris basin which first brought this important fact to light. The Eocene mammals, like the Carboniferous amphibians, the Mesozoic reptiles, and the Cretaceous forests, appear to spring full-grown from the earth, and this at nearly the same time in every part of the northern hemisphere. It has been suggested that they may have come in gradually without our knowledge in the Cretaceous period; but if so, we should have found some of their remains along with those of the Upper Cretaceous plants. But the prevalence of the great reptiles up to the close of the Cretaceous would seem to render the co-existence of large mammals unlikely. It has further been supposed that geological changes in the southern and northern hemispheres may have alternated with each other, so that there may be in the former Cretaceous beds in which the remains of ancestors of the Eocene mammals may be found. But we do not as yet know of such deposits. We may be content, therefore, to suppose that at the close of the Cretaceous there was established somewhere a sort of Eden for the first placental mammals, in which they were introduced and could live unharmed by the decaying monsters of the reptilian age, until the time came when they could increase and multiply and replenish the earth. The nearest approach to such a centre of mammalian life is perhaps to be found in those great American lake basins embedded in the mountains of the West, which have been so well described by Hayden and

Newberry, and which have yielded so many animal remains to the researches of Leidy, Marsh, and Cope.

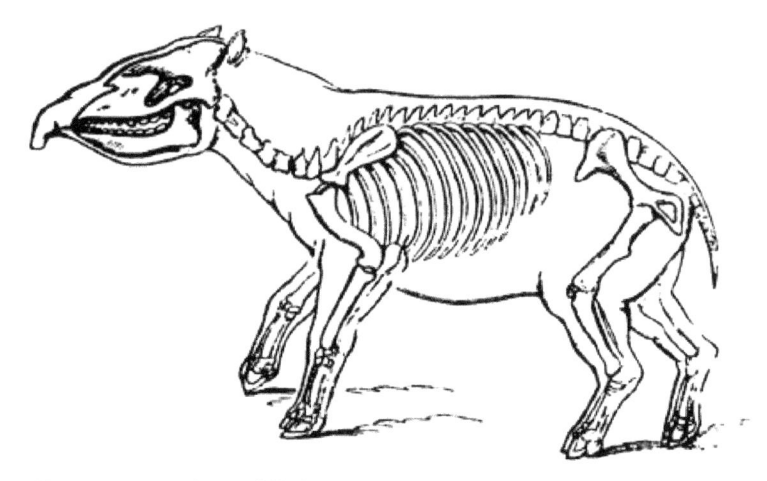

FIG. 170.—Restoration of *Palæotherium magnum*. Eocene.—After Cuvier and Owen.

The typical deposits of the Early Eocene have long been those of the Basin of Paris, where thick and highly fossiliferous deposits of this age rest on the more or less denuded surface of the Upper Chalk, and have afforded a rich harvest of remains of about fifty species of placental quadrupeds, whose bones have been found in the gypsum quarries of Montmartre. The great majority belong to the Ungulates, or hoofed animals, and the most abundant genera are those called by Cuvier *Palæotherium* (Fig. 170) and *Anoplotherium*, of which there are several species, and which have affinities with the modern Tapirs on the one hand, and with the Horse on the other. Of the Unguiculate or clawed orders there are carnivorous forms allied to the Hyæna and the Fox, a Bat and a Squirrel; and the Marsupials are represented by an Opossum. Lyell describes a bed of clay associated with the gypsum, in which are numerous footprints, probably produced on the margin of a lake. Many of these might be referred to the Palæothere and its allies; but there are others belonging to quadrupeds yet unknown, and there are also tracks of tortoises, crocodiles, and lizards, and of a large wading bird. Such a bed, perhaps deposited on the margin of a salt lake, resorted to as a "lick" by herbivorous animals, and by the carnivorous species which preyed on them, is well fitted, by the thronging life which it indicates, to teach how little we can know of the actual number and variety of the old inhabitants of the earth.

In England, Eocene beds of the age of those of Paris, occupy the valley of the Thames and the Isle of Wight and neighbouring parts of Hants. They have afforded mammalian fossils similar to those of Paris, though less abundantly, but they are rich in remains of marine animals and of land plants.

Instead of describing the well-known animals of the French and English Tertiaries, from these Eocene deposits upwards, I shall shortly sketch the succession in America, as worked out by Marsh and Cope, with the aid of the admirable summary given by Gaudry of the present state of knowledge with reference to the sequence of mammalian life from its appearance in the Early Eocene up to the present time.[80]

Eocene mammals, especially those gigantic whale-like creatures called *Zeuglodon* (Fig. 180), have been found in Eastern North America, but the most remarkable discoveries have been made in the Western Territories, where vast numbers of bones are imbedded in certain ancient and widespread lacustrine beds. It may be well to premise here that though the division into Eocene, Miocene, and Pliocene is recognised in America as well as in Europe, the limits of these groups may not precisely correspond with those in the Old World. Still we have this certain point of departure, that the Eocene begins where the peculiar animals of the Cretaceous end, and that the drying up of the later Cretaceous sea and the establishment of the Eocene land were probably nearly contemporaneous in both continents. It is true, however, in animals as in plants, that in the successive periods of the Tertiary, America presents an older aspect than Europe, just as its modern fauna still contains such old forms as the opossum.

It would seem that as the mountain-ranges and table-lands of Western America emerged from the Cretaceous waters, they became clothed with Eocene forests and inhabited by Eocene mammals. But the waters, dammed up by surrounding ridges, formed large lake basins, which were drained only by the slow excavation of "cañons" as the land rose still higher. In the successive deposits formed in these lakes both by ordinary deposition of silt and by paroxysmal showers of volcanic ashes were entombed great numbers of the animals which fed on their banks. It appears that these deposits, which in some places are estimated at not less than 8000 feet in thickness, hold the remains of three successive faunas, differing materially from each other, and representing the Lower, Middle, and Upper Eocene. On the flanks of the elevated region supporting the beds formed in the Eocene lakes, are other later lake basins of Miocene age, also abounding in animal remains. East of the Rocky Mountains, and also on the Pacific coast, are still later Pliocene deposits holding other and more modern Mammalia. The vast area of these formations and the complete sequence which they show are scarcely equalled elsewhere.

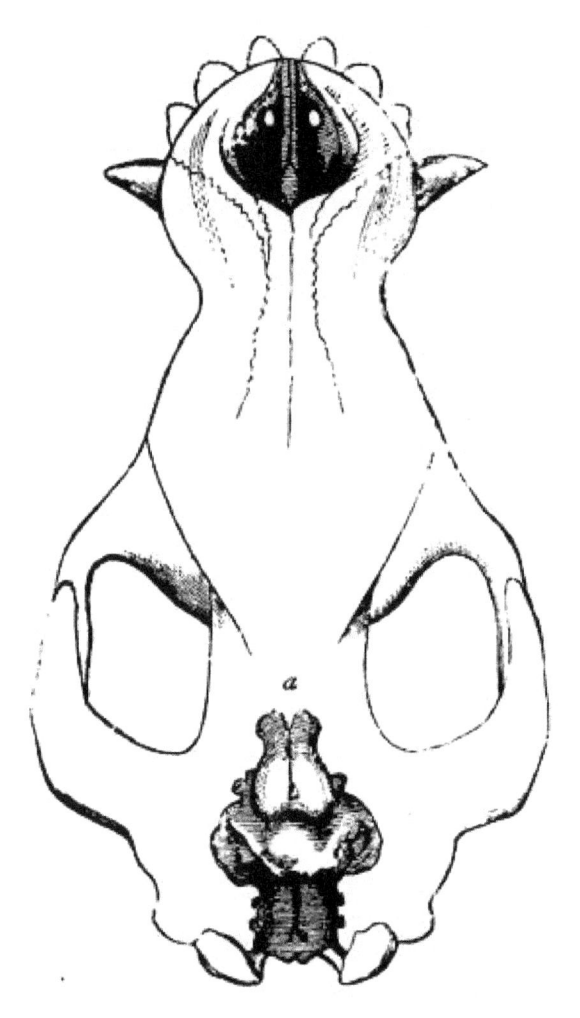

FIG. 171.—*Coryphodon Hamatus*. A Lower Eocene Perissodactyl skull, greatly reduced, showing small size of brain, *a*.—After Marsh.

As in the Paris basin, the large Ungulates constitute the most conspicuous feature. The great group is now usually divided into those that are odd-toed (Perissodactyl) and those that are even-toed (Artiodactyl). Though these are apparently arbitrary characters, they correspond with other more fundamental differences. The first includes such modern animals as the Rhinoceros, Tapir, and Horse. The second includes two somewhat distinct assemblages—that with mammillated teeth, of which the Hog and Hippopotamus are types (Bunodonts), and that with crescental plates of enamel in the teeth, of which the Ruminants like the Deer, Ox and Camel, are examples (Selenodonts).

FIG. 172.—Fore-foot of *Coryphodon*. Greatly reduced.—After Marsh.

The most characteristic animals of the lowest Eocene belong to the genus *Coryphodon* (Figs. 171, 172), which so abounded in Eocene America that bones of about 150 individuals were found by the Wheeler Expedition in one year in the Eocene beds of New Mexico. These animals in their dentition approached the American tapirs, except that they had great canines like the bear, while their feet resembled those of the elephant, and some of them attained the dimensions of the ox. *Coryphodon* is thus, as might be expected in a primal placental mammal, a creature of somewhat generalised type. Another point in which it resembles some at least of its early Tertiary contemporaries is the small size of the brain, especially in those parts of it supposed to minister to the intelligence and higher instincts (Fig. 171, *a*). It is certainly remarkable that as Tertiary time went on the successive groups of mammals were gifted with brains of larger and larger size, fitting them for higher functions; and ultimately for associating with man. Animals thus low in development of brain were probably slow and sluggish and stubbornly ferocious, and dependent on brute force for subsistence and defence; and they would have been altogether unsuitable for domestication had they lived to the present time.

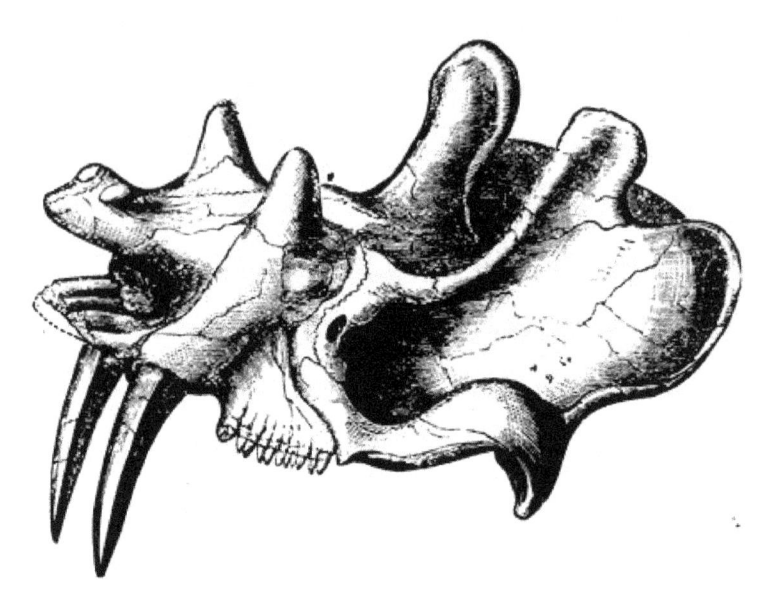

FIG. 173.—Skull of an Upper Eocene Perissodactyl (*Dinoceras mirabilis*), showing three pairs of horn-bases. Greatly reduced.—After Marsh.

In the Middle Eocene, the place of *Coryphodon* was taken by *Dinoceras* and allied forms. Some of the species nearly equalled the elephant in size, but had shorter and stouter limbs, each supported on five great toes—the most perfect possible sort of pedestal foot (Figs. 172, 174). They were heavily armed with immense canines on the upper jaws, and two or even three pairs of horns or hard protuberances on the head (Fig. 173). Creatures so supported and so armed, and living where food was plentiful, might well dispense with any great degree of intelligence, and their development of brain is consequently little better than that of *Coryphodon*. These great and characteristic Eocene families have no known successors; and in the Miocene age their place is taken by a very different group, that of which *Brontotherium* is the type (Fig. 175). They are creatures of huge size, with a pair of horn-cores on the nose, and feet with four toes in front and three behind, resembling in form those of the rhinoceros.

FIG. 174.—Fore-foot of *Dinoceras*. Greatly reduced.—After Marsh.

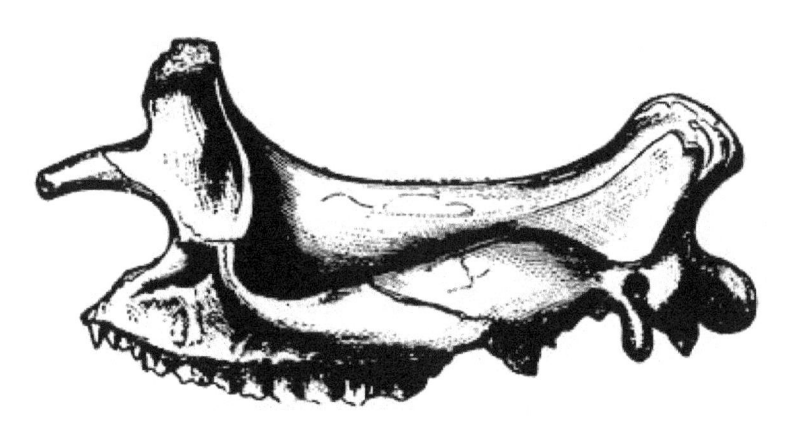

FIG. 175.—Skull of *Brontotherium ingens* (Marsh). Greatly reduced. A Miocene Perissodactyl.

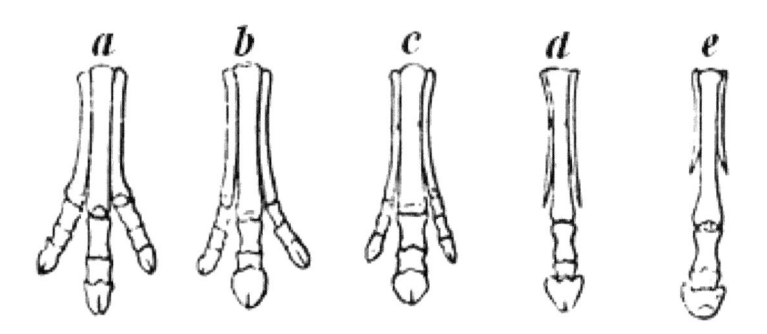

FIG. 176.—Series of Equine Feet.—After Marsh.

a, Orohippus, Eocene. *b, Miohippus*, Miocene. *c, Protohippus*, Lower Pliocene. *d, Pliohippus*, Upper Pliocene. *e, Equus*, Post-Pliocene and Modern.

While these gigantic Perissodactyles have no successors as yet known to us, another and less conspicuous Eocene type can be traced onward to modern times by a chain of successors which the imagination of evolutionists has converted into a veritable genetic series, to which they appeal as a "demonstration" of the process of descent with specific modifications. In the Lower Eocene are found the remains of a diminutive ungulate (*Eohippus*), of the stature of a moderately-sized dog. It has four toes and a rudiment of a fifth in front, and three toes behind; and has teeth slightly resembling those of the horse, but more simple and shorter in the crown. In this creature it has been supposed that we have a direct ancestor of the modern horse. A very similar genus (*Orohippus*), lacking only the fifth rudimentary toe, replaces *Eohippus* in the Middle Eocene. *Mesohippus* of the Lower Miocene is as large as a sheep, and has only three toes on the fore-foot and a splint bone, while its teeth assume a more equine character (Fig. 176). In the Upper Miocene *Miohippus* continues the line, while *Protohippus* of the Lower Pliocene is still more equine and as large as an ass, and corresponds with the European *Hipparion* in having the middle toe of each foot alone long enough to reach the ground. In the Upper Pliocene true horses appear with only a single toe, and splint bones instead of the others. In America, though the horse was unknown at the time of the discovery of the continent, several species occur in the Tertiary and Post-Pliocene, showing that the genus existed there up to a comparatively late period; and when re-introduced it has thriven and run wild in the more temperate regions. What cause could have led to its extinction in Post-Glacial times is as yet a mystery. This genealogy of the horse, independently of its evolutionist application, is very interesting. It shows that some Eocene types were suited to continuance, and even adapted for extension, while others were destined to become altogether extinct at an early date. It shows farther that the power of continuance resided not so much in the gigantic and prominent species as in smaller forms. It is to be observed, however, that Gaudry and other orthodox evolutionists in Europe deduce the horse, not from *Eohippus*, but from *Palæotherium*, and that it is equally impossible to verify either phylogeny, since the mere sequence of more or less closely allied species in time does not prove continuous derivation. Nor indeed are we certain that one-toed horses like those now living did not exist on the dry plains in Eocene times, since the inhabitants of these plains are probably unknown to us. An amusing illustration of the probable reason of the disappearance of the missing links has recently been given by a writer not very favourable to the new philosophy. The several consecutive species may be represented by coins. We may suppose, for example, sixpences to have been coined first, then sevenpenny and eightpenny pieces, and so on up to a shilling, then pieces representing thirteen, fourteen and fifteen pence, and so on up to a half-crown or crown; but all the intervening denominations between the sixpence and the shilling, and between the shilling and the half-crown, were found practically of little use. Hence few were coined, and they soon became obsolete. Thus the antiquary would find only a few denominations, and those connecting them would be seldom or never found. It is plain that if we could suppose that nations constructed their coinage after this unthinking and empirical fashion, and that if we were justified in ascribing a similar procedure to the Creator, it might help to account for the facts as we find them, otherwise we should rather suppose that in both cases something like plan and calculation determined the selection of the species produced, whether of coins or animals. But Chance is a blind goddess, and if we instal her as creator, we must expect the work to proceed by a series of abortive experiments.

The Perissodactyls are not numerous at present. The three groups represented by the Horse, Rhinoceros, and Tapir constitute the whole; and the two latter forms can be traced back to predecessors in Eocene times, even more closely resembling them than those supposed to be ancestors of the horse resemble that animal. But the few species now living have thus a vast surplusage of possible ancestors. Many species and genera are dropped without any modern representatives, so that the tendency has been to a gradual elimination of surplus types, until only a few isolated and somewhat specialised forms remain at present. Yet this process of elimination is not necessarily an evolution or survival of the fittest, in the sense of modern derivationists. It rather implies that in certain past states of the earth the conditions of life afforded scope for many forms not now required, or replaced by other types more suited to the advanced and specialised nature of the world.

On the other hand, the Artiodactyls have gained in numbers and importance, in comparison with their odd-toed comrades; and this, though an odd number, namely five, was the typical number with which the earliest quadrupedal forms began life far back in the Palæozoic. The typical Artiodactyls are those that cleave the hoof, and many of which also chew the cud; and they are of all others, the horse perhaps excepted, those that are most valuable to man. The lower type (Bunodont), to which the hog belongs, is the older; and many hog-like animals occur from the earlier Tertiary upwards. In the Upper Eocene, even-toed species appear with an approach at least to the crescent-shaped teeth of the modern deer and oxen. Some of the species are obviously forerunners of the modern antelopes and deer, though as yet destitute of horns or antlers. Others, like *Oreodon*, are of more hog-like aspect, though believed to have been ruminants (Fig. 177). These are characteristic of the Middle Miocene, at which stage true deer appear in Europe (*Dicroceras*), though they are not known in America until the Pliocene period. The earliest deer have small and simple antlers, these ornaments becoming larger and more elaborate in approaching the modern era. The hollow-horned ruminants appear for the first time in America in the Lower Pliocene; and no ancestry has so far been attempted to be traced for them. The antelopes of this group, as well as the gigantic *Sivatherium* of India,[81] allied to the modern prong-horned antelope of North America, were prominent in the Old World in the Miocene.

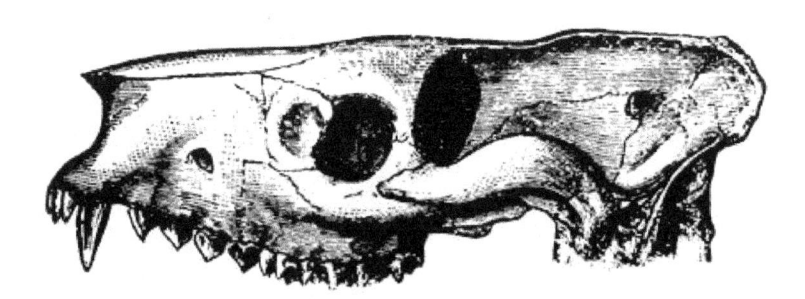

FIG. 177.—*Oreodon major*. A generalised Miocene ruminant, with affinities to the Deer, Camel, and Hog. Greatly reduced.—After Leidy.

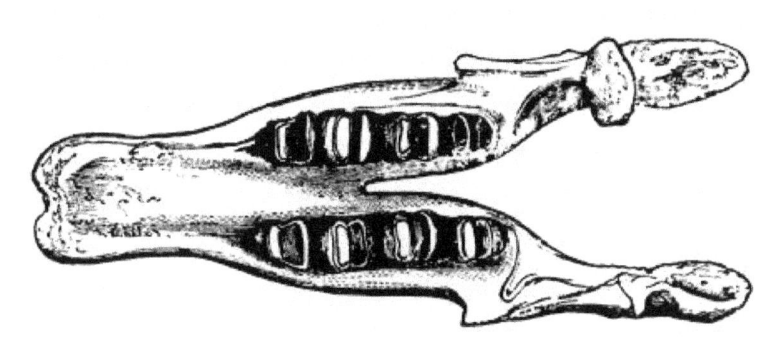

FIG. 178.—Lower Jaw of *Megatherium*. Greatly reduced. Post-Pliocene of South America.—After Owen.

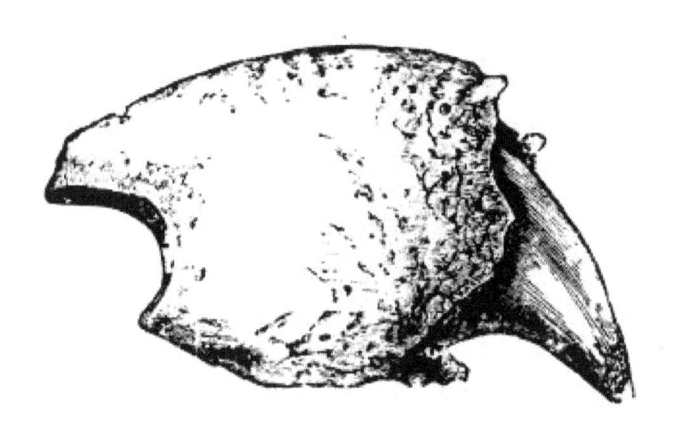

FIG. 179.—Ungual Phalanx and Claw-core of *Megatherium*. Greatly reduced.

A very noteworthy and specially American group of mammals is that of the *Edentates*, the Sloths and Ant-eaters, a group which *à priori* we should have supposed would have been one of the earliest in time. They appear, however, first in the Miocene, without even any suggested ancestry, and are represented from the first by large species, though they attain their grandest stature in the *Megatherium* and *Mylodon* of the Post-Pliocene (Figs. 178, 179), which were sloths of so gigantic size that they must have pulled down trees to feed on their leaves, unless, indeed, there were trees equally colossal for them to climb. But before the modern time, like the American horses, the larger herbivorous forms suddenly disappear, and are now represented only by a few diminutive South American species, which can scarcely, by any stretch of imagination, be supposed to be descendants of their gigantic predecessors. The history of these animals, like those of the great Tertiary marsupials of Australia and the many Miocene elephants of India, affords a remarkable illustration of the persistence of similar groups of creatures in successive ages in the same region, along with diminution in magnitude and number of species toward the modern times.

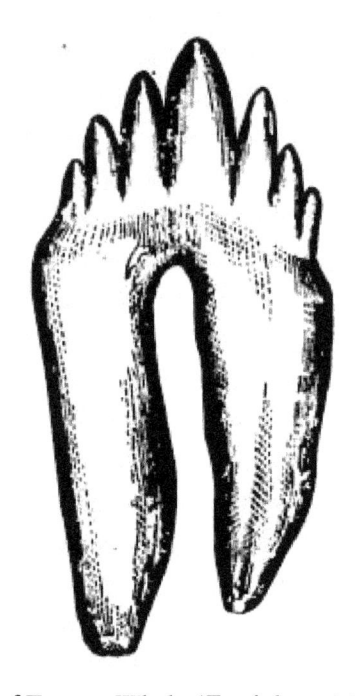

FIG. 180.—Tooth of Eocene Whale (*Zeuglodon cetioides*). One-half natural size.

The Whale-tribe (Cetaceans) at once in the earliest Eocene takes the place of the great Sea-lizards of the Cretaceous; and the oldest of the whales are in their dentition more perfect than any of their successors, since their teeth are each implanted by two roots, and have serrated crowns, like those of the Seals. The great Eocene whales of the Southern Atlantic (*Zeuglodon*) (Fig. 180), which have these characters, attained the length of seventy feet, and are undoubtedly the first of the whales in rank as well as in time. This is perhaps one of the most difficult facts to be explained on the theory of evolution. Allied to the whales is the small and peculiar group of the Sea-cows or Dugongs (*Sirenians*). These creatures, highly specialised and very distinct from all others, appear in the Early Tertiary in forms very similar to those which now exist, and probably in much more numerous species, and they pursue the even tenor of their way down to modern times without perceptible elevation or degradation. "We have questioned," says Gaudry, when speaking of the Tertiary Cetaceans, "these strange and gigantic sovereigns of the Tertiary oceans as to their progenitors—they leave us without reply." Their silence is the more significant as one can scarcely suppose these animals to have been nurtured in any limited or secluded space in the early stages of their development. The true Seals, which are more elevated than the Whales, and very different in type, appear much later, and without any probable ancestry.

The Elephants, two or three species of which constitute in the modern world the sole representatives of an order, are a remnant of an ancient race once vastly more numerous. They appear in Europe and Asia in the Miocene, when they were represented by three distinct genera (*Elephas*, *Mastodon*, and *Dinotherium*). The second genus (Fig. 181) differs from the proper Elephants in having tuberculated teeth, indicating a more swinish habit, and probably a more fierce disposition. The third (Fig. 182) is remarkable for the immense size of some of its species, far exceeding the modern Elephants, and has the farther peculiarity of a pair of descending tusks on the lower jaw, constituting a strong and heavy grubbing-hoe, with which it could probably dig deeply for roots. So important was the group in Miocene times that seven elephants are already known from this formation in India alone, besides three species of Mastodon. Four or five Miocene Mastodons are known in Europe, besides two *Dinotheria*; and the true Elephants appear there in the Pliocene, and continue to the beginning of the Modern. The elephantine animals are not known in America till the Pliocene, but in that and the Pleistocene, and perhaps up to the human period, the western continent, now altogether destitute of elephants, possessed several species both of *Elephas* and *Mastodon*, which extended, as in Siberia, even into the Arctic regions; and, as we know from specimens preserved in a frozen state in the latter region, some of the species were so protected by dense fur as to be able to endure extreme cold. The candid Gaudry closes his summary of the history and affinities of the elephantine animals with the words: "However, the sum of the differences compared with that of the resemblances is too great to permit us to indicate any relation of descent between the proboscidians and the animals of other orders known to us at present." So these greatest of all the animals of the land, with their strangely specialised forms and almost human sagacity, stand alone, without father or mother, without descent.

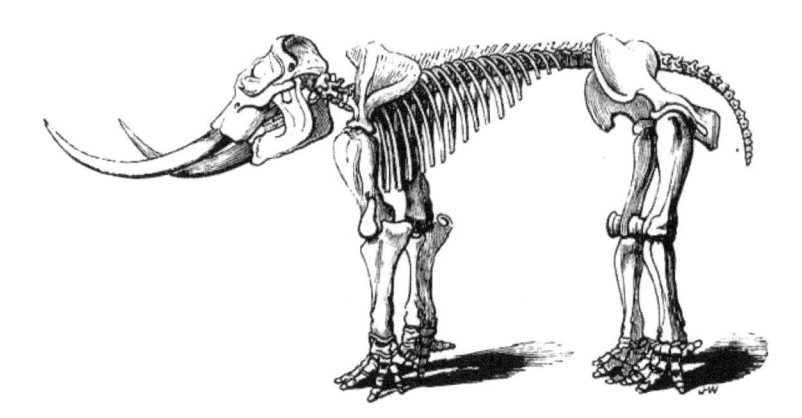

FIG. 181.—*Mastodon ohioticus.* An American Elephant. Post-Glacial.

FIG. 182.—Head of *Dinotherium giganteum.* Greatly reduced. Miocene of Europe.

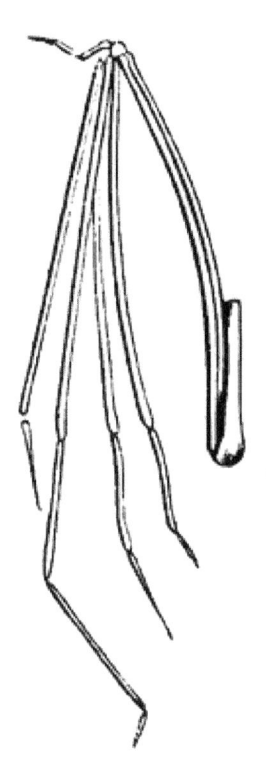

FIG. 183.—Wing of *Vespertilio aquensis*. An Eocene Bat. After Gaudry.

The Rodents, or gnawing animals, appear in the Early Eocene on both continents in familiar forms allied to our Squirrels and Rats. Porcupines and Beavers are added in the Miocene. This group seems thus to have continued much as it was; and it is still perhaps represented by as many species as at any previous time. Many of the ancient forms were, however, much larger than any modern species, and some of these larger forms[82] present singular points of approach to very distinct types, as, for example, to that of the Bears; but these large and composite species are long since extinct. The insectivorous mammals have much the same history with the Rodents. Such highly specialised and abnormal forms as the Bats might be supposed to be modern. But, strange to say, they appear with fully developed wings both in Europe and America in the Eocene (Fig. 183). Gaudry thinks that it is "natural to suppose" that there must have been species existing previously with shorter fingers and rudimentary wings; but there are no facts to support this supposition, which is the more questionable since the supposed rudimentary wings would be useless, and perhaps harmful to their possessors. Besides, if from the Eocene to the present the Bats have remained the same, how long would it take to develop an animal with ordinary feet, like those of a shrew, into a bat?

The Early Eocene was not altogether a time of peace in the animal world. The old carnivorous Saurians were dead and buried, but their place was taken by carnivorous mammals, allied to our modern Tigers, Hyænas, Foxes, and Weasels. The Carnivora, however, were subordinate in the Eocene, and, as already remarked, some of them appear to be intermediate between marsupial and placental forms—a fact which evolutionists have noticed with much satisfaction. They appear to attain to their culmination in the Miocene, when their powers seem to be proportionate to those of the great and well-armed quadrupeds they had to deal with. To this age belongs the introduction of the terrible "Cymetar-toothed Tiger" (*Machairodus*, Fig. 184). Its huge tusk-like canines and powerful limbs seem to fit it more than any other of the cat family for destructive efficiency. Yet ordinary cat-like animals were contemporary with it, and have survived it, since Machairodus disappears in the Post-Pliocene, though in previous periods it had been very widely distributed on both continents. It is a curious fact, perhaps of more significance in various ways than we yet understand, that the Dog-bear (*Arctocyon*), of the oldest French Eocene, believed to be the oldest placental mammal known, though technically placed among the Carnivora, has a kind of dentition indicating that, like the modern Bears, it was really omnivorous; and its skull shows some peculiarities tending to those of the Marsupials.

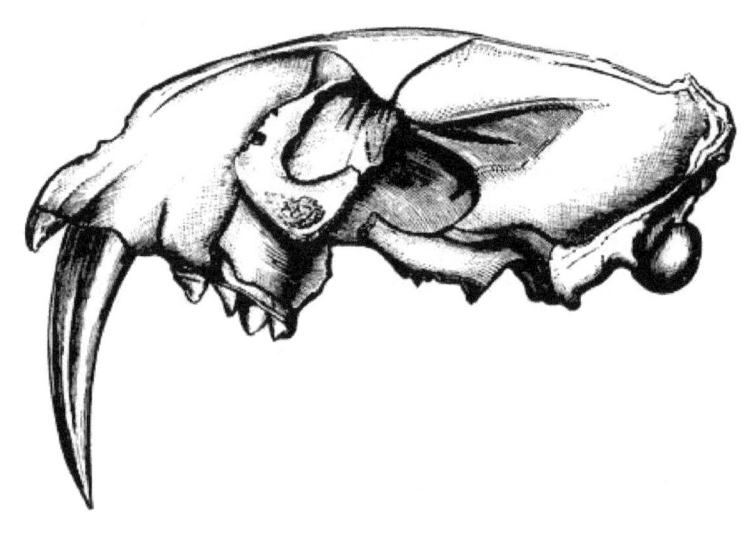

FIG. 184.—Skull of a Cymetar-toothed Tiger (*Machairodus cultridens*). Pliocene, France. Reduced.

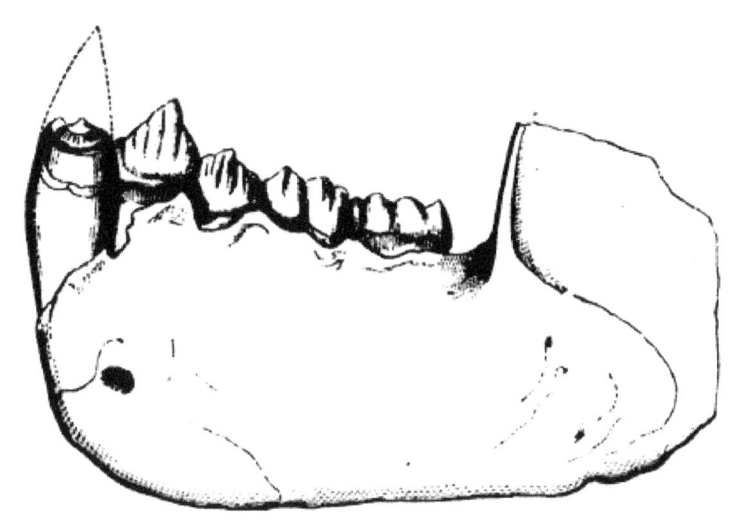

FIG. 185.—Lower Jaw of *Dryopithecus Fontani*. An Anthropoid Ape of the Middle Miocene of France. Natural size.

Much interest attaches to the first appearance of the order of Apes (*Quadrumana*), or, if we take the somewhat deceptive classification favoured by some modern zoologists, the *Primates*, including the apes and man. They begin in the Eocene, both in Europe and America, with the lowest tribe, that of the Lemurs, now confined to the island of Madagascar and parts of Africa and Southern Asia, and which may, Gaudry thinks, be modified Marsupials, though he admits that this is hard to understand. He mentions the resemblance of the teeth of monkeys to those of some hog-like animals, a resemblance, however, merely marking a similarity of food, and suggests on this ground that some of the primitive ancestors of the hog may have also given rise to the Monkeys. In the Miocene of Europe and Asia we have true Apes; and one of these, which rivals man in stature (*Dryopithecus*), belongs to the group of the gibbons, or long-armed apes, one of the higher families of the modern *Quadrumana* (Fig. 185). This animal presents, indeed, the nearest approach to man made by any Tertiary mammal. Still the differences are great, as, for instance, in the much larger size of the canines and premolars. Yet so much confidence has Gaudry in the resemblances, that he even ventures

to suggest that certain flint chips found in the Miocene of Thenay, and which have been supposed to indicate human workmanship, may have been chipped by the hands of *Dryopithecus*. Should this view be adopted by evolutionists, it will at least have the effect of preventing flint chips from being received as evidences of the antiquity of man.

It is scarcely necessary to sum up this review of the history of the Tertiary mammals. Much that has been said may be modified or changed by future discoveries; but the great facts of the late appearance of the placental mammals, of their rapid introduction, with their ordinal differentiation nearly complete over all the continents, of the speedy culmination and early decadence of many types, and of the unchanged permanence of others, must in the main be sustained. It is not too much to say that to account for these facts the evolutionist must abandon the idea of gradual change, and adopt that of "critical periods" when sudden changes occurred. The history becomes inexplicable, unless with Mivart, Le Conte, and Saporta, we admit "periods of rapid evolution" alternating with others of stagnation or retrogression; and if we admit these, we practically fall back on the old idea of creation; only it may perhaps be "Creation by Law."

CONTEMPORARIES OF POST-GLACIAL MAN. From a painting by Waterhouse Hawkins.

CHAPTER X.

THE ADVENT OF MAN.

HITHERTO we have met with no trace of man or of his works. Yet there have been in our upward progress from the dawn of life mute prophecies of his advent. Man is in his bodily frame a vertebrate animal and a mammal; and when first the Amphibians were introduced in the Palæozoic, the framework of man's body was already sketched out and its principles settled. Those great reptilian lords, the biped Saurians of the Mesozoic, already foreshadowed his erect posture, though their limbs may have been more ornithic than mammalian. The gradual advance in the brain-development of the Tertiary mammals presaged a coming time when mind would obtain the mastery over claw and tooth and horn; and in the Miocene ages there was already some hint of the precise style of structure in which this new creative idea would be realised. Yet it might have been impossible to imagine beforehand the vast changes which this new idea would inaugurate. In the lower animals such intelligence as they possess is so tied to the physical organisation that it manifests itself as a mechanical unvarying instinct. Man bursts this bond, and in doing so revolutionises the whole scheme of nature. Old things are now put to new uses, the face of nature is changed, varied arts are introduced, and thought enters into the domain of general and abstract truth. Objects are arranged, classified, understood, and while in some respects the whole creation is made to groan under the tyrannous inventions of man, yet these are the inventions of imagination and design. They are the triumph, not of brute force, but of will and intelligence.

That man was not in all the earlier ages of the world, except in these prophecies of his coming, geology assures us. That he is, we know. How he came to be, is, independently of Divine revelation, an impenetrable mystery—one which it is doubtful if in all its bearings science will ever be competent to solve. Yet there are legitimate scientific questions of great interest relating to the time and manner of his appearance, and to the condition of his earlier existence and subsequent history, which belong to geology, and in which so great stores of material have been accumulated that a treatise rather than a chapter would be required for their discussion. We may endeavour to select a few of the more important points.

One of the first questions meeting us is that which relates to the point in geological time signalised by the advent of our species. In the Eocene period our continents were being gradually raised out of the ocean, and were still in great part under the waters, which several times returned upon the land, and seemed ready again to engulf it. In this period not only have we no traces of man, but all the higher animals of that age are now extinct. In the later

Eocene and Miocene the extent of land became greater, but it was so disposed as to allow the influx into the Arctic Sea of vast volumes of heated water from the equatorial regions; and there may have been also astronomical causes at work to increase this influx of warm water, and so to raise the temperature of the Arctic regions still higher.[83] The middle period of the Tertiary was undoubtedly a time very favourable to the wide distribution of the higher forms of life both animal and vegetable. But we cannot trace man or any of the contemporary mammals back to the Miocene. In the Pliocene the continents had attained to their present elevations, and climates were not dissimilar from those prevailing at present; but still we have no certain indication of the presence of man; and if other modern mammals extend back to this period their number is very small. In this age also the greater part of the continents must have been covered with a great thickness of soil and disintegrated rock favourable to vegetation, and there seemed nothing to preclude the introduction of man. But a new and at first sight most unfavourable change was to intervene. Whether through internal changes affecting the distribution of land and water, or through astronomical vicissitudes, the northern hemisphere, and possibly the whole world, entered on an era of refrigeration, the so-called "Glacial Age" of the Post-Pliocene or Pleistocene period. That in this period our continents as far south as the latitude of 40° were overwhelmed with ice or ice-laden seas is rendered evident by the fact that the whole surface up to several thousands of feet above the sea-level has been bared of its accumulated *débris* and polished and grooved by ice, and laden with boulders and other glacial deposits, while in many places at heights of even 1,000 or 1,200 feet these deposits contain seashells of species now living in the colder parts of the ocean. These phenomena do not exist in the tropical regions, except in the vicinity of high mountains, but they recur in the southern hemisphere. It is still uncertain whether the period of greatest cold in the two hemispheres was at the same time or in successive ages. Geologically, however, they are approximately contemporaneous, both occurring between the end of the Pliocene and the modern period; but nevertheless they may not have coincided in absolute date.

Very different views have been held as to the precise condition of the continents in the Glacial Age, though all agree in the prevalence of cold and the action of ice, and in the fact of a great submergence at one or more stages of the period. My own conclusions, which I have advocated elsewhere,[84] and which are based on extensive study of the northern parts of America, where the deposits of this age are more widely developed than elsewhere, are that there was one great subsidence, leading to a condition in which the lower levels of the continents were covered with ice-laden water and the higher regions were occupied with permanent snow and glaciers. This submergence went on till even high mountains 4,000 feet or more in elevation were under water. Then there was a gradual though intermittent elevation, during which the climate became ameliorated, and lastly there was a condition in which the land of the northern hemisphere stood higher than at present, and which immediately preceded the modern period. As these conditions have great significance in relation to the appearance of man, I have tabulated them for reference as they occur in Scandinavia, Great Britain, and North America. The so-called "Interglacial Periods" of some geologists are in reality local results of the stages of intermittent elevation in which were deposited beds which in some cases, as in Scotland, Sweden, and Eastern Canada, hold sea-shells, and in others, as in the central areas of North America, contain remains of plants of northern species.

We shall name, for convenience, the parts of this Pleistocene revolution which include the great subsidence and glaciation, the *Glacial* Age, that extending from the re-elevation to the modern the *Post-glacial*.

The Glacial Age proved fatal to a large proportion of the land life of the previous periods. According to Professor Boyd Dawkins, out of fifty-three species known in Britain in the Post-glacial, only twelve are survivors of the Pliocene; and probably the proportions would not be greater in any part of the northern hemisphere. Some, however, did survive, either by migrating southward or by being inhabitants of places less severely affected than most by the general cold and submergence. There was thus no absolute break in the chain of life effected by the Glacial Age.

TABLE OF PLEISTOCENE DEPOSITS IN SCANDINAVIA, ENGLAND, AND AMERICA. (Order descending.)

SCANDINAVIA. (Torell.)	GREAT BRITAIN. (Lyell, &c.)	NORTH AMERICA.
Valley-clays and Heath-sands of Sweden. (No fossils.)	Hoxne Deposits and Upper Terrace Gravels. Palæolithic Implements.	Terrace Gravels and Loess Deposits.
Terrace-gravels of Norway and Sweden. (No fossils.)	Upper Glacial Beds. Bridlington Beds. Upper Boulder Beds.	Placer Gravels of West.
		Do. Sand and Gravel, Newer Boulder Drift.

Dryas-clay with Fossil plants of northern species.	So-called "Interglacial" Deposits.	So-called Interglacial Beds, with Plants, &c. Loess Deposits of Mississippi.
Uddevalla beds with Boreal Marine shells.	Clyde Beds and Marine Clays.	Upper Leda Clay and Champlain Clay, with Boreal Shells. White Silts of British Columbia.
	Mid-Glacial Sands.	Erie Clays and similar Beds of West.
Yoldia Clay and Sand. Arctic Marine Shells.		Lower Leda Clay, with Arctic Shells.
Yellow Stony Clay and Sand, and Gravel of Scania.		Port Hudson Deposit of Mississippi.
		"Syrtensian" Beds of New Brunswick.
		Orange Sand of Mississippi.
"Moraines de Fond," or Boulder Clay proper.	Till, or Older Boulder Clay.	Boulder Clays, with Local and some Travelled Boulders.
Ancient Diluvial Sand.	Pebbly Beds and Weyburne Sands, Lignitic Forest Beds.	Old Land Surfaces-- Peat under Boulder Clay, Local Gravels and Sands.
		Pre-glacial Gravels of British Columbia.

In what part of this sequence did man appear? In answer to this, I think it is now generally admitted that he is not certainly known earlier than the Post-glacial period. Various supposed indications of his presence in "Inter-glacial"

Glacial, Pliocene, and even Miocene deposits have proved on examination to be unreliable. America has recently put forth claims to have been inhabited by man in the Pliocene, on the faith of remains found in auriferous gravels in the West. But the facts that the implements and bones found are modern in type, that the gravels were deeply mined by the Indians, and that the objects found, as mortars for dressing gravel, etc., are in many cases such as they would be likely to leave in their excavations, have discredited these supposed discoveries. Still more recently, chipped flints found in gravels in New Jersey, by Abbott, have been supposed to carry back the Indians of the East coast to the Glacial period. It is evident, however, from the description of these deposits by the late Mr. Belt and by Professor Cook, director of the Survey of New Jersey, that they are really Post-glacial, that their age must be estimated by study of the local conditions, and that there is no good ground for correlating them with the upper members of the true Glacial drift to the northwards, with which they had been somewhat rashly identified. Irrespective of the doubtful character of many if not all of the so-called implements, the deposits in which they are found is confessedly not a product of the ice of the Glacial period proper, whether that was, as some maintain, a period of land glaciation as far south as New Jersey or not. It belongs to a time of denudation by water, aided perhaps by floating ice, and is not necessarily older than the river gravels of the Somme, which, like it, contain boulders and imply conditions of torrential action and climate which have long since passed away. If, however, these implements are genuine, they would imply the presence of Palæocosmic or Antediluvian man in America. This would in itself be an important discovery.

For the present, therefore, man is geologically a Post-glacial species, and there is nothing unreasonable in supposing that he dates no farther back, since several animals his contemporaries are in the same case; and by supposing him to have originated after the Glacial age we avoid the difficulties attendant on his survival of that great revolution. The only necessity for supposing an earlier appearance arises from the requirements of the hypothesis of evolution. Those, however, who hold this theory, may with Haeckel take refuge in that shadowy continent supposed to have extended from Africa to Australia,[85] and to have sheltered man in his transition from the ape to humanity, in the Tertiary period. The name Lemuria is taken from the Lemurs, supposed ancestors of the Apes, which still haunt the margin of the Indian Ocean; but it may be taken also in its old Latin sense of ghosts of the evil dead; and as we are not likely to obtain any more tangible evidence of the old natives of Lemuria, perhaps we may hope that some spiritualist may succeed in charming them from the vasty deep for our enlightenment. Should this be so, it is to be hoped that no "drum ecclesiastic" will be beaten to drive them away till they have revealed all they can tell.

It may be well to add that, in addition to the negative evidence, there is at least one positive evidence of the recent origin of man which has been well urged by Le Conte. It is this: animals have continued long in geological time in the inverse ratio of their rank. Some Mesozoic protozoa still survive. So do many early Tertiary mollusks. But the mammals are of much less duration. No living species goes back farther than the Pliocene. Few extend farther than the Glacial age. On the same principle it is not to be expected that man, the highest of all animals, should extend far back in geological time.

Accepting the Post-glacial age as that of the advent of man, it may be interesting to ask what we know of the condition of our continents when he appeared. In Western Asia, in Europe, except in its more northern portions, and it would now seem also in America, man had been introduced at a time closely following the emergence of the land from the Glacial sea. At this time the land area of both continents was larger than it is at present, and the character of the fauna shows that much of the surface was occupied with great steppes or prairies, over which migration would be easy; while there were probably connections by land or chains of islands between the continents of the northern hemisphere. The land animals of the continents were more numerous and of greater stature than at present. Several species of elephants (Fig. 186) and a rhinoceros roamed over the plains. The formidable *Elasmotherium* (Fig. 187),[86] an animal allied to the rhinoceros, but more fleet and active, and of immense size, inhabited Asia and Europe. Hippopotami, wild horses, the gigantic Irish stag, several species of wild cattle, and bisons of greater size than their successors, haunted the streams and steppes. The cave bear, the cave lion, the spotted hyæna, and possibly the *Machairodus*, were among the beasts of prey even in the temperate latitudes. The climate must have been a continental one, ranging through considerable extremes; but the conditions favoured migration of animals on the great scale, so as to avoid these extremes, and hence species of types now comparatively restricted enjoyed a wide distribution.

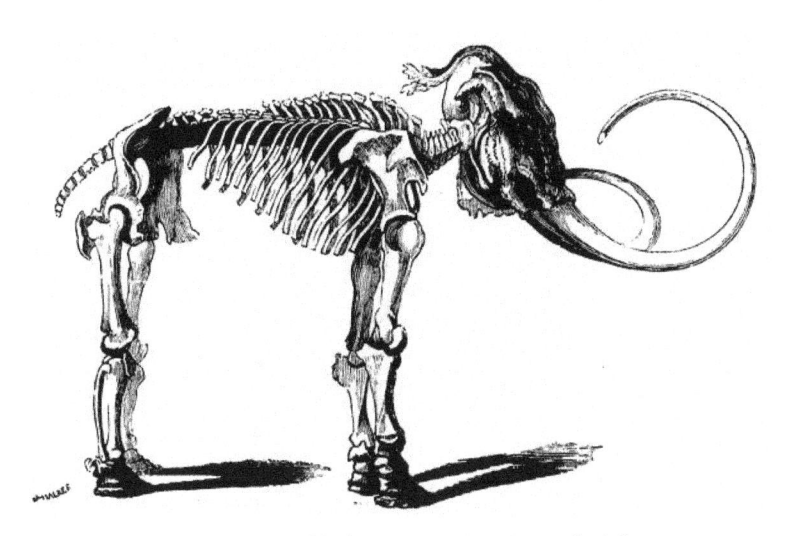

FIG. 186.—*Elephas primigenius.* Post-glacial.

To establish themselves in such a world, the primitive men must have been no puny race, either in mind or body, and they must have been sheltered in some Eden of plenty and comparative safety till, by increase of numbers, invention of weapons and implements, and domestication of useful animals, they became able to cope with the monarchs of the waste. But this position once attained in the original seats of the species, the wide continents presented great facilities for their movements, and there were ample stores of food for wandering tribes subsisting by the chase.

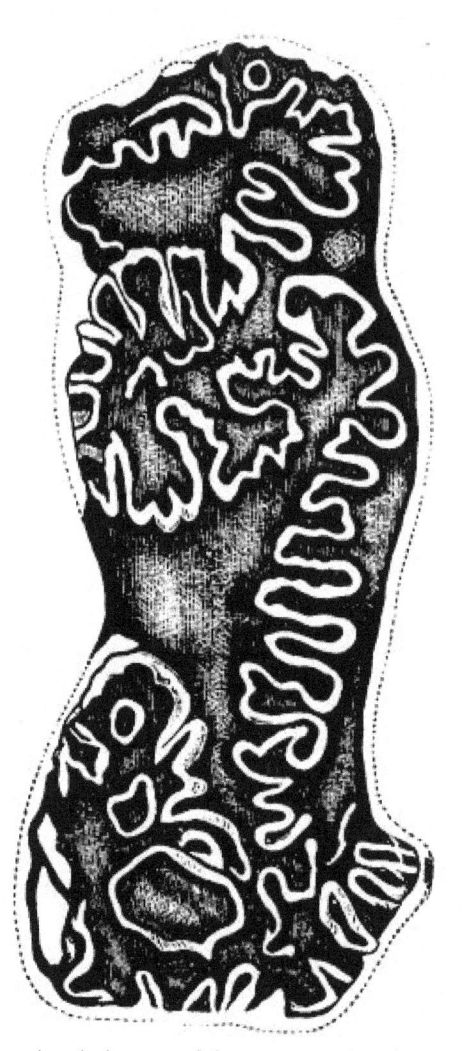

With such views the skeletons of the most ancient known men[87] fully accord. They indicate

FIG. 187.—Tooth of *Elasmotherium*. Grinding surface, natural size. Siberia. From *Nature*. a people of great stature, of powerful muscular development, especially in the lower limbs, of large brain, indicating great capacity and resources (Fig. 188), but of coarse Turanian features, like those of the tribes that now roam over the plains of Northern Asia (Fig. 189). They used flint and bone implements, which they manufactured with much skill (Figs. 190, 191). They were probably clothed in dressed skins, ornamented with embroidery, in the manner of the North American Indians. They used shells and carved bones as ornaments. Recent discoveries at Soloutre, in France, render it probable that some of the tribes had tamed the horse, and resided in fortified villages. They buried their dead with offerings, indicating a belief in immortality. These Post-glacial men are certainly known as yet only in Europe and Western Asia; and we cannot therefore determine if they represent the average man of the period. There were in Belgium and other parts of Europe, men of smaller stature and of lower cranial type, contemporary or nearly so with the higher race. There may have been fruit-eating or agricultural peoples in the more genial and fertile lands of the east and south. The conditions above sketched are, I think, fairly deducible from the facts stated by Christy and Lartet, Dupont, Rivière, Dawkins, and others, who have studied the remains of these early men, the Palæolithic men of some writers, or the men of the Mammoth age, and whom I have elsewhere

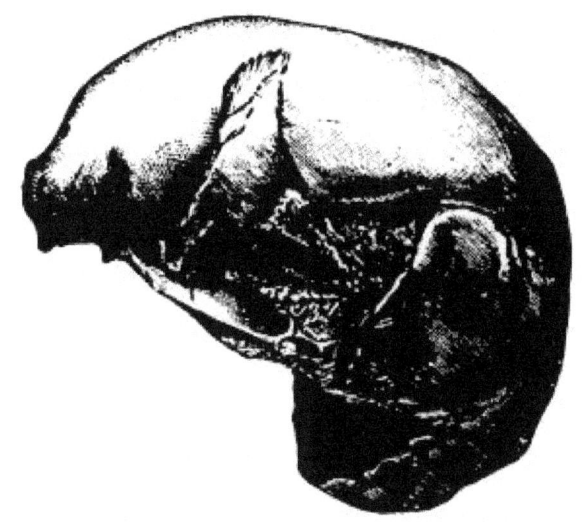

named Palæocosmic men, as a term less objectionable than those founded on implements not confined to any age, or animals which may have long antedated FIG. 188.—Engis Skull. Reduced.—After Lyell. The Skull of one of the Men of the Mammoth age. man. Recent discoveries in the caves of Spy in Belgium,[88] taken in connection with the previous discoveries of Schmerling and Dupont, seem to show the existence in that country of men of the low-browed Neanderthal or Canstadt type (Fig. 189 third outline), perhaps locally preceding but perhaps contemporaneous with, the larger and better developed men of the Cro-Magnon type (Fig. 189 first outline). These two types are, however, allied, and there are intermediate forms, so that they are to be regarded as two races of Palæocosmic men not more dissimilar than we find in cognate rude races at present.

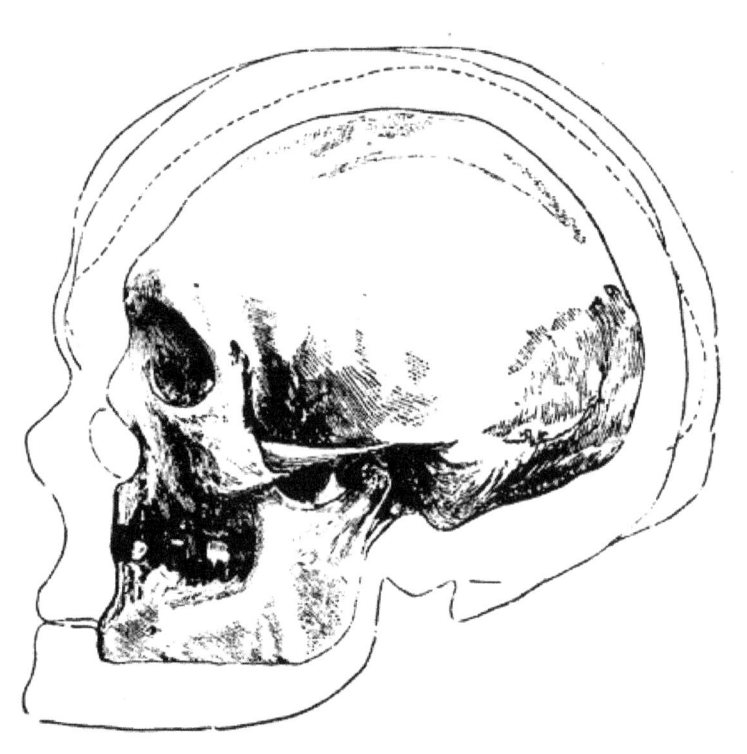

FIG. 189.—Outlines of Three Prehistoric European Skulls compared with an American Skull from Hochelaga.

Outer outline, Cro-Magnon Skull. Second outline, Engis Skull. Third outline (dotted), Neanderthal Skull. Inner figure, Hochelagan Skull on a smaller scale.

They were succeeded in Western Europe by a smaller and less elevated race, identical apparently with the modern Lapps and Basques, and in whose time the mammoth and many large animals had disappeared, Europe had become clad with dense forests, and the reindeer had extended his range far to the south, while the land of our continents had become narrowed to its present limits, or even less. The cause of these changes must have been physical, and to some extent cataclysmal; and its wide-spread and effectual character is shown by the fact that it exterminated so many animals of both continents which had survived the Glacial age. Similar testimony is borne by the occurrence of the implements and remains of Palæocosmic men in gravels and in diluvial clays in caverns, and by the changes of level and deep erosions of valleys that are referable to the close of the Palæocosmic age. The most probable agencies in this revolution were subsidences of the land, accompanied with climatal changes; but the precise nature and extent of these is still unknown; and the prevalent tendency on the part of geologists to stretch the doctrine of uniformity, so valuable within proper limits, to the absurd extreme of excluding all changes not exemplified even in amount in the modern period, will probably for some time prevent any adequate conception of them.

It would be premature to correlate what is yet known of the Palæocosmic age with historical periods; but the tendency of the facts accumulated is, I think, toward the identification of the Palæocosmic men with the Antediluvians; and their Neocosmic successors, whether of the reindeer age, of the Danish shell-mounds, or the Swiss lake habitations, with Postdiluvian and still existing tribes.

FIG. 190.—Flint Implement found in Kent's Cavern, Torquay, under four feet of cave mud and one foot of stalagmite.—After Pengelly.

After what has been already said, it will be unnecessary to dwell upon the characteristics of the first race of men known to us. They were rude and uncivilised, in so far as outward appliances are concerned; but they are confessedly altogether men, and in no respect akin to apes, and their volume of brain is rather greater than that of the average European of to-day; so that they must have had quite as much natural sagacity and capacity for culture, and, like the modern and historic Turanian nations, they were probably superior to the average European in the instinct for art and construction. Thus if we suppose these men derived from apes by any process of gradual change, we must look for their brute ancestors, not in the Pliocene or Miocene, but in the Eocene itself. This causes us to recur to the doctrine of critical periods, when many species were introduced together, alternating with periods of decay and extinction. Post-glacial man appears at the end of a time of sifting and trial, a time in which a vast number of species succumbed to great physical reverses. No very great number of species came in with him, and in the early period of his history there was a decadence or destruction either by the diluvial cataclysm or gradually. Out of ninety-eight species of mammals contemporary with early man in Europe, forty-one FIG. 191.— Bone Harpoon (Palæocosmic), from Périgord Cavern. are wholly or locally extinct, and none have been introduced except those brought by man

himself. Thus man stands alone, the grand product of his period and a lord of creation, for whom great preparatory changes were made, and multitudes of lower animals swept away to make room for him. According to our sacred Scriptures, this change is still imperfect, and great additional ameliorations would have taken place but for a moral catastrophe not within the domain of geology—the fall of man. If we identify the Palæocosmic men with the Antediluvians of the same venerable record, the roving tribes whose remains are known to us represent that part of the race of Cain of whom Jubal was the father, the nomads dwelling in tents, as distinguished from the settled agricultural peoples. In this case, also, the catastrophe which destroyed these rude and lawless men was that which culminated in the deluge of Noah, which may represent the extinction of the last great body of this primitive race, whose arts, handed down to the physically inferior men of Postdiluvian times, astonish us by their early development in Chaldæa and in Egypt.

If man is so recent geologically, he may still be very old historically; and the question remains, Have we any facts bearing on the absolute antiquity of man? For the properly historical aspect of this question, I may refer to the excellent work of Canon Rawlinson on the *Origin of Nations*,[89] which shows conclusively that the historic origin of all the great nations of antiquity extends backward less than 4,400 years from our time. Beyond this we have, however, the Palæocosmic or Antediluvian men; and their extension backward seems limited geologically only by the close of the Glacial period, while many hold that the Genealogy in Genesis does not require us to limit very narrowly their antiquity. The date of the Glacial period is, however, at present very uncertain. On the one hand, some geologists, like Lyell, have supposed it may be as far back as 200,000 years ago. Others, like Croll, are contented with the more moderate estimate of 80,000 years. On the other hand, the calculations of Andrews, based on the recession of the American lakes, those of Winchell on the recession of the Falls of St. Anthony, and the recent surveys of the recession of the Falls of Niagara, reduce the time to from 7,000 to 10,000 years. It is impossible in the present state of knowledge to settle these disputes; but one may refer in the sequel to some of the evidences which have been adduced in favour of great antiquity. Since the publication of the second edition of this work Prof. Prestwich, in a paper read before the Geological Society,[90] has brought forward other reasons which induce him to conclude that the close of the Glacial epoch occurred "from 8,000 to 10,000 years since." It is true, he admits, on geological evidence still in dispute, that man may have existed in Europe before that time, and he also admits, on historical, not geological evidence, the existence of "Neolithic" man in Asia, "at an earlier date than 4,000 B.C." Still the repudiation, by so good an authority, of the exaggerated antiquity which it has been the fashion, since the rise of Darwinian evolution, to assign to man, contrary to the geological evidence, is a satisfactory indication of a return to more rational views; and when geologists get rid of the fiction of a continental ice-sheet, still farther progress in this direction may be expected.

We may, I think, at once take it for granted, that none of the Neocosmic races date farther back than the origin of the great eastern nations. There are certainly no geological evidences requiring a greater antiquity, for in their time the land had attained to its present configuration, and the changes which have occurred in the succession of forests and the growth of peat are such as our experience in America shows to be possibly quite modern. There is besides no doubt that these people, from the Reindeer men of France and Belgium to the people of the Swiss lakes, are modern races, whose descendants still live in Europe. We can thus limit our inquiry to the Palæocosmic men; and with respect to them we know only what may be gathered from a consideration of the physical changes which have occurred since they lived.

In Europe a great number of considerations have been adduced as evidence of their high antiquity; and these deserve careful attention, though I think it will be found that they are all liable to serious objections or great abatements on geological grounds.

(1) The occurrence of human remains with those of animals now extinct affords no certain evidence of antiquity. Admitting that human remains are found along with those of the mammoth in Europe, and with those of the mastodon in America, the question remains, How late did these species survive? In Europe we know that several large animals now extinct existed up to comparatively modern times. This is the case with the Irish deer (*Megaceros*), the urus, the aurochs, and the reindeer, in temperate Europe. How long previously the mammoth or the hairy rhinoceros disappeared we do not know, but need not suppose the time very long.

(2) The accumulation of sediment or of stalagmite over human remains in caverns is not necessarily indicative of very great antiquity. We know that in favourable circumstances mud, sand, and gravel may be rapidly deposited in caves by land floods or river inundations, and that *débris* of various sorts accumulates in such places from decay of rock and vegetable and animal agencies. The deposition of stalagmite is also very variable in its rate; and the fact that it is being very slowly deposited in any cave now does not prove that more rapid deposition may not have taken place formerly. Dawkins and others have ascertained a rate of a quarter of an inch per annum in some caverns; and this would allow the stalagmite crust of Kent's cave, for which an antiquity of half a million of years has been claimed, to have been formed in a thousand years.

FIG. 192.—Sketch of a Mammoth, carved on a portion of a Tusk of the same Animal (Lartet).

(3) The erosion of river valleys to great depths since the Glacial period fails to establish the great antiquity of the caves left on their sides or the high level gravels of their banks. Throughout the northern hemisphere, the river valleys are of old date, and were merely filled with loose detritus in the Glacial age. The sweeping out of this *débris* would be a rapid process, more especially when changes of level were occurring, and when the rainfall was greater than at present. Besides, as Croll has well remarked, the actual configuration of our continents, the amount of drift still remaining, and the imperfect manner in which the river valleys have been cleared out, all testify to the comparative recency of the Glacial period.[91] These considerations would, indeed, materially reduce the antiquity which he claims on astronomical grounds for the ice age.

(4) The growth of peat and the deposition of silt are very deceptive as indications of great antiquity. For instance, accurate observations made by a French engineer in the construction of docks at St. Nazaire,[92] show that in 1,600 years the Loire had deposited over Gallo-Roman remains six metres of mud. Relics of the Bronze age occur below these at a depth indicating 500 years previously as their date; and the beginning of the modern deposit of the Loire would, on the same evidence, be only 6,000 years ago. Hilgard's observations on the delta of the Mississippi in like manner tend greatly to reduce our estimates of the time occupied in the deposit of the modern silt of that river. The peat deposit at Abbeville, at the mouth of the Somme, has been supposed to have required 30,000 years for its formation. But this estimate was based upon the present rate of growth; and, as Andrews has shown, the fact admitted by Boucher de Perthes, that birch stems three feet high stand in this peat, implies a much more rapid rate, which is also proved by the depth at which Roman remains have been found. In like manner the Scandinavian peats, to which a fabulous antiquity has been ascribed, have

been proved to be comparatively modern by the depths at which metallic works of art are found in them.

(5) The paucity of remains of Palæocosmic men in Europe, with their wide distribution, indicate that their sojourn was not long, or that the population was very small and much scattered. Even in a few thousands of years, an active and vigorous people, living in a country well supplied with food, must have multiplied greatly, and must have left considerable remains. On the theory that these men inhabited Europe even for 2,000 years, we have to suppose that the greater part of their remains have been swept away, or remain under the waters, or buried out of sight in diluvial sediments.

(6) Much importance has been attached to the early works and high culture of Egypt and Chaldæa, as evidence of vast time during which arts were growing from a supposed rude stone age. But it must be observed that no such period is known to antedate civilisation in the East, and that if the early empires were established by survivors of the Deluge, they must have brought with them the culture of Antediluvian times. Farther, the notion of men emerging from a half-brutal state, and from the use of the rudest implements, is purely conjectural and not supported by facts. In America, where the semi-civilised agricultural races are unquestionably the oldest, the rudest possible implements were used by these partially civilised agricultural people along with polished stone and metal; and Schliemann has shown that a rude stone age succeeded the civilisation of Troy, and this at a time when Phœnicia and Egypt were at the height of their civilisation. Such facts, which might fill volumes, show how little value is to be attached to supposed ages of rough and polished stone.

(7) The difficulties attending the establishment of geological dates for deposits like those containing the remains of men are very great. They are altogether superficial and local, not widespread marine beds in which a distinct order of superposition can be clearly traced. They are not easily separated from the glacial beds below, or from those above which have been modified by human agency, by land-floods, or by landslips. Thus the application of geological criteria of age to them is very difficult and uncertain. Evidence of this could easily be given, in the many errors which have been promulgated, and which have had to be retracted by their authors, or have been disproved by the observations of others. For example, no country was at one time richer in supposed evidences of the antiquity of man than Scandinavia; but Professor Torell, the director of the Geological Survey of Sweden, has recently made a careful re-examination of the facts, and has found that there is no evidence whatever of the existence of man in Scandinavia before the Neolithic or polished stone age. There are, however, evidences of considerable changes of level since that time, and it would seem even since the twelfth century of our era. The remarkable and seemingly

inexcusable errors of observation referred to in Professor Torell's memoir, should enforce a caution on geologists as to the uncertainties of such evidence. Lyell sifted the testimony bearing on this subject with great care in the first edition of his *Antiquity of Man*. In later editions he had to make large abatements, and now much of the evidence in the latest edition would have to be withdrawn or otherwise applied.

From all these considerations the conclusion is obvious that while we have no certain data for assigning a definite number of years to the residence of man on the earth, we have no geological evidence for the rash assertion often made that in comparison with historical periods the date of the earliest races of men recedes into a dim, mysterious, and measureless antiquity. On the basis of that Lyellian principle of the application of modern causes to explain past changes, which is the stable foundation of modern geology, we fail to erect any such edifice as the indefinite antiquity of man, or to extend this comparatively insignificant interval to an equality with the long æons of the preceding Tertiary. The demand for such indefinite extension of the history of man rests not on geological facts, but on the necessities of hypotheses which, whatever their foundation, have no basis in the discoveries of that science, and are not required to account for the sequence which it discloses.

CHAPTER XI.

REVIEW OF THE HISTORY OF LIFE.

WHAT general conclusions can we reach as to this long and strange history of the progress of life on our planet? Perhaps the most comprehensive of these is that the links in the chain of life, or rather in its many chains, are not scattered and disunited things, but members of a great and complex plan; and that when we discern their combinations and their pattern, we find them not only orderly and symmetrical, but all tending to one point and bound to one central object, even the throne of the Eternal. It must also appear evident that the original plan of nature, both in the animal and vegetable worlds, was too vast to be realised at one time on a globe so limited as ours, but had to be distributed in time as well as in space, thus realising the idea of time-worlds: successive æons in which, one after the other, the work of creation could rise to successive stages of perfection and completeness till it culminated in man. All this is sufficiently plain on the theistic view of nature, and may suffice for those who reverently regard the God of nature as the Father of their own spirits. But there are others who ask further questions. Do we know anything of the secondary causes and origin of life, of the manner of its introduction and advance, of the laws of its succession?

As to the first of these questions, it is certain that, up to this time, the origination of the living being from the non-living is an inscrutable mystery. No one has witnessed this change, or has been able to effect it experimentally. Nor have we any direct evidence of the origination of one specific type from another. Such reasonings as assume the possibility of these things, or on analogical grounds assert their probability, belong rather to the domain of philosophical speculation than to science. As to the laws of the succession of life, however, it is possible to learn something from the sequence of facts as already ascertained; and though much remains to be discovered, there are a few leading statements on this subject which can already be made with safety.

Unity and uniformity, within the limits imposed by progress and increasing complexity, can be affirmed of the whole process. From the dawn of life to the present time the great laws of physical nature which operate on animals and plants have been uniform. These stable laws have regulated the action of the outer world on organisms. The plans of structure of these organisms laid down at the first have been followed throughout. Thus the succession of life presents nothing fortuitous or arbitrary, but a continuous plan carried out uniformly in time and space, with certain materials of fixed properties, and with certain structures predetermined from the first. There is, for example, a great sameness of plan throughout the whole history of the marine invertebrate life of the Palæozoic. If we turn over the pages of an illustrated text-book of geology, or examine the cases or drawers of a collection of fossils, we shall find extending through every succeeding formation, representative forms of Crustaceans, Mollusks, and Corals, in such a manner as to indicate that in each successive period there has been a reproduction of the same type with modifications; and if the series is not continuous, this appears to be due to lack of specimens, or to abrupt physical changes; since sometimes, where two formations pass into each other, we find a gradual change in the fossils by the dropping out and introduction of species one by one. Thus in the whole of the great Palæozoic period, both in its fauna and flora, we have a continuity and similarity of a most marked character.

There is, indeed, nothing to preclude the supposition that many forms reckoned as species are really only race modifications. My own provisional conclusion, based on the study of Palæozoic plants, published many years ago,[93] is that the general law will be found to be the existence of distinct specific types independent of each other, but liable in geological time to minor modifications, which have often been regarded as distinct species.

While this unity of successive faunæ at first sight presents an appearance of hereditary succession, it loses much of this character when we consider the number of new types introduced without apparent predecessors, or ceasing without successors, and the almost changeless persistence of other types; the necessity that there should be similarity of type in successive faunæ on any hypothesis of a continuous plan; and, above all, the fact that the recurrence of representative species or races in large proportion marks times of decadence rather than of expansion in the types to which they belong. To return to a later period, this is very manifest in that singular resemblance which obtains between the modern mammals of South America and Australia and their immediate fossil predecessors—the phenomenon being here manifestly that of decadence of large and abundant species into a few depauperated representatives. This will be found to be a very general law, elevation being accompanied by the abrupt appearance of new types, and decadence by the apparent continuation of old species, or modifications of them.

This resemblance with difference in successive faunæ also connects itself very directly with the successive elevations and depressions of our continental plateaus in geological time.

Every great Palæozoic limestone, for example, indicates a depression with succeeding elevation. On each elevation marine animals were driven back into the ocean, and on each depression swarmed in over the land, reinforced by new species, either then introduced or derived by migration from other localities. In like manner on every depression, land plants and animals were driven in upon insular areas, and on re-elevation again spread themselves widely. Now I think it will be found to be a law here that periods of expansion were eminently those of introduction of new specific types, and periods of contraction those of extinction, and also of continuance of old types under new varietal forms. It must also be borne in mind that all the leading types of invertebrate life were early introduced, that change within these was necessarily limited, and that elevation could take place mainly by the introduction of the vertebrate orders. So in plants, Cryptogams early attained their maximum as well as Gymnosperms, and elevation occurred in the introduction of Phænogams.

Another allied fact is the simultaneous appearance of like types of life in one and the same geological period, over widely separated regions of the earth's surface. This strikes us especially in the comparatively simple and homogeneous life-dynasties of the Palæozoic, when for example we find the same types of Silurian Graptolites, Trilobites and Brachiopods appearing simultaneously in Australia, America, and Europe. Perhaps in no department is it more impressive than in the introduction in the Devonian and Carboniferous ages of that grand cryptogamous and gymnospermous flora which ranges from Brazil to Spitzbergen, and from Australia to Scotland, accompanied in all by the same groups of marine invertebrates; or in the like wholesale production of modern types of trees in the Cretaceous. Such facts may depend either on that long life of specific types which gives them ample time to spread to all possible habitats, before their extinction; or on some general law whereby the conditions suitable to similar types of life emerge at one time in all parts of the world. Both causes may be influential, as the one does not exclude the other, and there is reason to believe that both are natural facts. Should it be ultimately proved that species allied and representative, but distinct in origin, come into being simultaneously everywhere, we shall arrive at one of the laws of creation, and one probably connected with the gradual change of the physical conditions of the world.

A closely related truth is the periodicity of introduction of species. They come in by bursts or flood-tides at particular points of time, while these great life-waves are followed and preceded by times of ebb in which little that is new is being produced. We labour in our investigation of this matter under the disadvantage that the modern period is evidently one of the times of pause in the creative work. Had our time been that of the early Tertiary or early Mesozoic, our views as to the question of origin of species might have been very different. It is a striking fact, in illustration of this, that since the Glacial age no new species of mammal can be proved to have originated on our continents, while a great number of large and conspicuous forms have disappeared. It is possible that the proximate or secondary causes of the ebb and flow of life-production may be in part at least physical; but other and more important efficient causes may be behind these. In any case these undulations in the history of life are in harmony with much that we see in other departments of nature.

It results from the above and the immediately preceding statement that specific and generic types enter on the stage in great force, and gradually taper off toward extinction. They should so appear in the geological diagrams made to illustrate the succession of life. This applies even to those forms of life which come in with fewest species and under the most humble guise. What a remarkable swarming, for example, there must have been of Marsupial Mammals in the early Mesozoic; and in the Coal formation the only known Pulmonates, four or five in number, belong to as many generic types.

I have already referred to the permanence of certain species in geological time. I may now place this in connection with the law of origination and more or less continuous transmission of varietal forms. I may, perhaps, best illustrate this in connection with a group of species with which I am very familiar, that which came into our seas at the beginning of the Glacial age, and still exists. With regard to their permanence, it can be affirmed that the shells now elevated in Wales to 1,200 and in Canada to 600 feet above the sea, and which lived before the last great revolution of our continents, a period vastly remote as compared with human history, differ in no tittle from their modern successors after thousands or tens of thousands of generations. It can also be affirmed that the more variable species appear under precisely the same varietal forms then as now, though these varieties have changed much in their local distribution. The real import of these statements, which might also be made with regard to other groups well known to palæontologists, is of so great significance that it can be realised only after we have thought of the vast time and numerous changes through which these humble creatures have survived. I may call in evidence here a familiar British and American animal, the common sand clam, *Mya arenaria*, and its relative, *Mya truncata*, which now inhabit together all the northern seas; for the Pacific specimens, from Japan and California, though differently named, are undoubtedly the same. *Mya truncata* appears in Europe in the older Pliocene, and was followed by *M. arenaria* a little later. Both shells occur in the Pleistocene of America, and their several varietal forms had then already developed themselves, and remain the same to-day; so that these humble mollusks, littoral in their habits, and subjected to a great variety of conditions, have continued, perhaps for one or two thousand centuries, to construct their shells precisely as at present. Nor are there any indications of a transition between the two species. Similar statements may be made with regard to other mollusks of the Pliocene and Modern periods, and there are even species which extend unchanged from the early Eocene. Nor is it impossible that some modern bivalves of the Brachiopod group may be scarcely modified descendants even of Palæozoic species.

Perhaps some of the most remarkable facts in connection with the permanence of species and varietal forms are those furnished by that magnificent flora which burst in all its majesty on the American continent in the Cretaceous period, and still survives among us even in some of its specific types, I say survives; for we have but a remnant of its forms living, and comparatively little that is new has probably been added since. Take, for example, the facts stated in Chapter VIII. as to the continuance to the present time of species of plants introduced in the Cretaceous and Eocene, and which thus came in at the very time when the great Mesozoic reptiles were decaying or had just disappeared, and when the placental mammals were being introduced. Some of these plants must have propagated themselves unchanged for half a million of years.

Plants and the lower tribes of animals are, however, more permanent than the higher animals; and a strange contrast is afforded to the foregoing examples of persistence by the repeated revolutions that have affected vertebrate life since the Mesozoic age. Yet even in the case of vertebrates there seems to have been little change, except in the extinction of species, since the Pliocene period.

In conclusion of this review, can we formulate a few of the general laws, or perhaps I had better call them the general conclusions respecting life, in which all palæontologists may agree? Perhaps it is not possible to do this at present satisfactorily, but the attempt may do no harm. We may, then, I think, make the following affirmations:—

(1) The existence of life and organisation on the earth is not eternal, or even coeval with the beginning of the physical universe, but may possibly date from Laurentian or immediately pre-Laurentian times.

(2) The introduction of new species of animals and plants has been a continuous process, not necessarily in the sense of derivation of one species from another, but in the higher sense of the continued operation of the cause or causes which introduced life at first. This, as already stated, I take to be the true theological or Scriptural as well as scientific idea of what we ordinarily and somewhat loosely term creation.

(3) Though thus continuous, the process has not been uniform; but periods of rapid production of species have alternated with others in which many disappeared and few were introduced. This may have been an effect of physical cycles reacting on the progress of life.

(4) Species, like individuals, have greater energy and vitality in their younger stages, and rapidly assume all their varietal forms, and extend themselves as widely as external circumstances will permit. Like individuals, also, they have their periods of old age and decay, though the life of some species has been of enormous duration in comparison with that of others; the difference appearing to be connected with degrees of adaptation to different conditions of life.

(5) Many allied species, constituting groups of animals and plants, have made their appearance at once in various parts of the earth, and these groups have obeyed the same laws with the individual and the species in culminating rapidly, and then slowly diminishing, though a large group once introduced has rarely disappeared altogether.

(6) Groups of species, as genera and orders, do not usually begin with their highest or lowest forms, but with intermediate and generalised types, and they show a capacity for both elevation and degradation in their subsequent history.

(7) The history of life presents a progress from the lower to the higher, and from the simpler to the more complex, and from the more generalised to the more specialised. In this progress new types are introduced, and take the place of the older ones, which sink to a relatively subordinate place, and become thus degraded. But the physical and organic changes have been so correlated and adjusted that life has not only always maintained its existence, but has been enabled to assume more complex forms, and that older forms have been made to prepare the way for newer, so that there has been on the whole a steady elevation culminating in man himself. Elevation and specialisation have, however, been secured at the expense of vital energy and range of adaptation, until the new element of a rational and inventive nature was introduced in the case of man.

(8) In regard to the larger and more distinct types, we cannot find evidence that they have, in their introduction, been preceded by similar forms connecting them with previous groups; but there is reason to believe that many supposed representative species in successive formations are really only races or varieties.

(9) In so far as we can trace their history, specific types are permanent in their characters from their introduction to their extinction, and their earlier varietal forms are similar to their later ones.

(10) Palæontology furnishes no direct evidence, perhaps never can furnish any, as to the actual transformation of one species into another, or as to the actual circumstances of creation of a species, but the drift of its testimony is

to show that species come in *per saltum*, rather than by any slow and gradual process.

(11) The origin and history of life cannot, any more than the origin and determination of matter and force, be explained on purely material grounds, but involve the consideration of power referable to the unseen and spiritual world.

Different minds may state these principles in different ways, but I believe that in so far as palæontology is concerned, in substance they must hold good, at least as steps to higher truths. And now I may be permitted to add that we should be thankful that it is given to us to deal with so great questions, and that in doing so deep humility, earnest seeking for truth, patient collection of all facts, self-denying abstinence from hasty generalisations, forbearance and generous estimation with regard to our fellow-labourers, and reliance on that Divine Spirit which has breathed into us our intelligent life, and is the source of all true wisdom, are the qualities which best become us.

As we have traced onward the succession of life, reference has been made here and there to the defects of those bold theories of descent with modification which are held forth in our time as the true bond of the links of the chain of life. It must have been apparent that these theories, however specious when placed in connection with a limited induction of facts selected for the purpose of illustrating them, are very far from affording a satisfactory solution of all difficulties. They cannot perhaps be expected to take us back to the origin of living beings; but they also fail to explain why so vast numbers of highly organised species struggle into existence simultaneously in one age and disappear in another, why no continuous chain of succession in time can be found gradually blending species into each other, and why in the natural succession of things degradation under the influence of external conditions and final extinction seem to be laws of organic existence. It is useless here to appeal to the imperfection of the record or to the movements or migrations of species. The record is now in many important parts too complete, and the simultaneousness of the entrance of the faunas and floras too certainly established, while the moving of species from place to place only evades the difficulty. The truth is that such hypotheses are at present premature, and that we require to have larger collections of facts. Independently of this, however, it would seem that from a philosophical point of view all theories of evolution, as at present applied to life, are fundamentally defective in being too partial in their character; and this applies more particularly to those which are "monstic" or "agnostic," and thus endeavour to dispense with a Creative Will behind nature. It may be instructive to illustrate from the facts developed in preceding chapters this feature of most of the attempts at generalisation on this subject.

First, then, these hypotheses are too partial, in their tendency to refer numerous and complex phenomena to one cause, or to a few causes only, when all trustworthy analogy would indicate that they must result from many concurrent forces and determinations of force. We have of late been very familiar with those ingenious, not to say amusing, speculations in which some entomologists and botanists have indulged with reference to the mutual relations of flowers and haustellate insects. Geologically the facts oblige us to begin with Cryptogamous plants and mandibulate insects; and out of the desire of insects for non-existent honey, and the adaptations of plants to the requirements of non-existent suctorial apparatus, we have to evolve the marvellous complexity of floral form and colouring, and the exquisitely delicate apparatus of the mouths of haustellate insects. Now when it is borne in mind that this theory implies a mental confusion on our part precisely similar to that which in the department of mechanics actuates the seekers for perpetual motion, that we have not the smallest tittle of evidence that the changes required have actually occurred in any one case, and that the thousands of other structures and relations of the plant and the insect have to be worked out by a series of concurrent evolutions so complex and absolutely incalculable in the aggregate that the cycles and epicycles of the Ptolemaic astronomy were child's play in comparison, we need not wonder that the common sense of mankind revolts against such fancies, and that we are accused of attempting to construct the universe by methods that would baffle Omnipotence itself, because they are simply absurd. In this aspect of them, indeed, such speculations are necessarily futile, because no mind can grasp all the complexities of even any one case, and it is useless to follow out an imaginary line of development which unexplained facts must contradict at every step. This is also no doubt the reason why all recent attempts at constructing "Phylogenies" are so changeable, and why no two experts can agree about almost any of them.

A second aspect in which such speculations are too partial is in the unwarranted use which they make of analogy. It is not unusual to find such analogies as that between the embryonic development of the individual animal and the succession of animals in geological time placed on a level with that reasoning from analogy by which geologists apply modern causes to explain geological formations. No claim could be more unfounded. When the geologist studies ancient limestones built up of the remains of corals, and then applies the phenomena of modern coral reefs to explain their origin, he brings the latter to bear on the former by an analogy which includes not merely the apparent results but the causes at work, and the conditions of their action; and it is on this that the validity of his comparison depends, in so far as it relates to similarity of mode of formation. But when we compare the development of an animal from an embryo cell with the progress of animals in time, though we have a curious analogy as to the steps of the process, the conditions and agents at work are known to be altogether dissimilar, and therefore we have no evidence whatever as to identity of cause, and our reasoning becomes at once the most transparent of fallacies. Farther, we have no right here to overlook the fact that the conditions of the embryo are determined by those of a previous adult, and that no sooner does this hereditary potentiality produce a new adult animal than the terrible external agencies of the physical world, in presence of which all life exists, begin to tell on the organism, and after a struggle of longer or shorter duration it succumbs to death, and its substance returns into inorganic nature, a law from which even the longer life of the species does not seem to exempt it. All this is so plain and manifest that it is extraordinary that evolutionists will continue to use such partial and imperfect arguments. Another example may be taken from that application of the doctrine of natural selection to explain the introduction of species in geological time which is so elaborately discussed by Sir C. Lyell in the last edition of his *Principles of Geology*. The great geologist evidently leans strongly to the theory, and claims for it the "highest degree of probability," yet he perceives that there is a serious gap in it; since no modern fact has ever proved the origin of a new species by modification. Such a gap, if it existed in those grand analogies by which we explain geological formations through modern causes, would be admitted to be fatal.

A third illustration of the partial character of these hypotheses may be taken from the use made of the theory deduced from modern physical discoveries, that life must be merely a product of the continuous operation of physical laws. The assumption, for it is nothing more, that the phenomena of life are produced merely by some arrangement of physical forces, even if it be admitted to be true, gives only a partial explanation of the possible origin of life. It does not account for the fact that life as a force or combination of forces is set in antagonism to all other forces. It does not account for the marvellous connection of life with organisation. It does not account for the determination and arrangement of forces implied in life. A very simple illustration may make this plain. If the problem to be solved were the origin of the mariner's compass, one might assert that it is wholly a physical arrangement both as to matter and force. Another might assert that it involves mind and intelligence in addition. In some sense both would be right. The properties of magnetic force and of iron or steel are purely physical, and it might even be within the bounds of possibility that somewhere in the universe a mass of natural loadstone may have been so balanced as to swing in harmony with the earth's magnetism. Yet we should surely be regarded as very credulous if we could be induced to believe that the mariner's compass has originated in that way. This argument applies with a thousandfold greater force to the origin of life, which involves even in its simplest forms so many more adjustments of force and so much more complex machinery.

Fourthly, these hypotheses are partial, inasmuch as they fail to account for the vastly varied and correlated interdependencies of natural things and forces, and for the unity of plan which pervades the whole. These can be explained only by taking into the account another element from without. Even when it professes to admit the existence of a God, the evolutionist reasoning of our day limits itself practically to the physical or visible universe, and leaves entirely out of sight the power of the unseen and spiritual, as if this were something with which science has nothing to do, but which belongs only to imagination or sentiment. So much has this been the case that when recently a few physicists and naturalists have turned to this aspect of the subject, they have seemed to be teaching new and startling truths, though only reviving some of the oldest and most permanent ideas of our race. From the dawn of human thought it has been the conclusion alike of philosophers, theologians, and the common sense of mankind, that the seen can be explained only by reference to the unseen, and that any merely physical theory of the world is necessarily partial. This, too, is the position of our sacred Scriptures, and is broadly stated in their opening verse; and indeed it lies alike at the basis of all true religion and all sound philosophy, for it must necessarily be that "the things that are seen are temporal, the things that are unseen, eternal." With reference to the primal aggregation of energy in the visible universe, with reference to the introduction of life, with reference to the soul of man, with reference to the heavenly gifts of genius and prophecy, with reference to the introduction of the Saviour Himself into the world, and with reference to the spiritual gifts and graces of God's people, all these spring not from sporadic acts of intervention, but from the continuous action of God and the unseen world; and this, we must never forget, is the true ideal of creation in Scripture and in sound theology. Only in such exceptional and little influential philosophies as that of Democritus, and in the speculations of a few men carried off their balance by the brilliant physical discoveries of our age, has this necessarily partial and imperfect view been adopted. Never indeed was its imperfection more clear than in the light of modern science.

Geology, by tracing back all present things to their origin, was the first science to establish on a basis of observed facts the necessity of a beginning and end of the world. But even physical science now teaches us that the visible universe is a vast machine for the dissipation of energy; that the processes going on in it must have had a beginning in time, and that all things tend to a final and helpless equilibrium. This necessity implies an unseen power, an invisible universe, in which the visible universe must have originated, and to which its energy is ever returning. The hiatus between the seen and the unseen may be bridged over by the conceptions of atomic vortices of force, and by the universal and continuous ether; but whether or not, it has become clear that the conception of the unseen as existing has become necessary to our belief in the possible existence of the physical universe itself, even without taking life into the account.

It is in the domain of life, however, that this necessity becomes most apparent; and it is in the plant that we first clearly perceive a visible testimony to that unseen which is the counterpart of the seen. Life in the plant opposes the outward rush of force in our system, arrests a part of it on its way, fixes it as potential energy, and thus, forming a mere eddy, so to speak, in the process of dissipation of energy, it accumulates that on which animal life and man himself may subsist, and assert for a time supremacy over the seen and temporal on behalf of the unseen and eternal. I say, for a time, because life is, in the visible universe, as at present constituted, but a temporary exception, introduced from that unseen world where it is no longer the exception but the eternal rule. In a still higher sense, then, than that in which matter and force testify to a Creator, organisation and life, whether in the plant, the animal, or man, bear the same testimony, and exist as outposts put forth in the succession of ages from that higher heaven that surrounds the visible universe. In them, as in dead matter, Almighty power is no doubt conditioned by law, yet they bear more distinctly upon them the impress of their Maker, and while all explanations of the physical universe which refuse to recognise its spiritual and unseen origin must necessarily be partial and in the end incomprehensible, this destiny falls more quickly and surely on the attempt to account for life and its succession on merely materialistic principles.

Here, however, we must remember that creation, as maintained against such materialistic evolution, whether by theology, philosophy, or Holy Scripture, is necessarily a continuous, nay, an eternal influence, not an intervention of disconnected acts. It is the true continuity, which includes and binds together all other continuity.

It is here that natural science meets with theology, not as an antagonist, but as a friend and ally in its time of greatest need; and I must here record my belief that neither men of science nor theologians have a right to separate what God in Holy Scripture has joined together, or to build up a wall between nature and religion, and write upon it "no thoroughfare." The science that does this must be impotent to explain nature and without hold on the higher sentiments of man. The theology that does this must sink into mere superstition.

In the light of all these considerations, whether bearing on our knowledge or our ignorance, a higher and deeper question presents itself, namely, that as to the relation of nature and of man to a Personal Creator. To this it seems to me that the study of the succession of life yields no uncertain reply. Call the progress of life an evolution if you will; trace it back to primæval Protozoa, or to a congeries of atoms: still the truth remains that nothing can be evolved out of these primitive materials except what they originally contained. Now we find in the existence of man, and in the tendency of the scheme of nature towards his introduction, evidence that at least all that is involved in the reasoning and moral nature of man must have existed potentially before atoms began to shape themselves into crystals or into organic forms. Nay, more than this is implied, for we do not know that man and what he has hitherto been and done constitute the ultimate perfection of nature, and we must suspect that something much more than what we see in man must be required for the origination of the chain of life. What does this prove, in any sense in which human reason can understand it? Nothing less, it seems to me, than that doctrine of the Almighty Divine Logos, or Creative Reason, as the cause of all things, asserted in our sacred Scriptures, and held in one form or another by all the greatest thinkers who have attempted to deal with the question of origins. Falling back on this great truth, whether presented to us in the simple "God said" of Genesis, or in the more definite form of the New Testament, "The Word was with God, and the Word was God," we find ourselves in the presence of a Divine plan pervading all the ages of the earth's history and culminating in man, who presents for the first time the image and likeness of the Divine Maker; and this forms the true nexus of all the separate chains of life. Had man never existed, such reasoning might have been speculative merely, but the existence of man, taken in connection with the progress of the plan which has terminated in his advent, proves the existence of God.

Divine revelation carries us a step farther, and teaches us to recognise in Jesus of Nazareth God manifest in the flesh, the Divine Logos dwelling among men. But though this is a doctrine of revelation and not of science, it is in perfect harmony with the plan of progress which we have been sketching. It is the natural outcome of a process leading to the introduction of a rational and accountable being, understanding something of the works and ways of God, that to him God should reveal Himself, and that the Divine Logos, by whom were "constituted the ages"[94] of the world's geological history, should preside also over its future consummation, when all the degradation that has sprung from the aberrations of fallen and imperfect humanity shall be removed, and man himself shall become fully a partaker of the Divine nature.

The world we live in is thus not necessarily a finished world, and it is now marred by the sins of man. What it may be in the future, we can perhaps as little guess as an intelligence studying the Palæozoic world could have understood that of the present time. But it is a glorious truth to know that our Maker has revealed Himself to us also as a Saviour, and that as individuals we shall not perish, to be replaced by an improved species in the future, but that we ourselves, as sons of God, may enter into and possess the new earth and new heavens of future æons of the universe. Thus it would seem that the Gospel of Jesus Christ is that which was wanting to complete and justify the history of nature by bringing to light the final "restitution of all things," and our own union to God in a happy immortality.

FOOTNOTES:

1 Croll has elaborated this calculation in his work, *Climate and Time.*

2 Sept. 1879.

3 Analyses recently made by Mr. C. Hoffman, of the Geological Survey of Canada, show that beds of graphitic gneiss, some of them 8 feet in thickness, contain as much as 25·5 to 30 per cent. of carbon, the remaining earthy matter consisting principally of silica, alumina, and lime. The graphite from veins was nearly pure carbon, containing from 97·6 to 99·8 per cent. of that substance.

4 Sometimes separated as a distinct order under the name of *Radiolaria.*

5 *Loftusia Columbiana*, Dawson, from British Columbia, is the only Carboniferous species yet known.

6 See Nicholson in the *Memoirs* of the Palæontographical Society.

7 *Archæospherinæ* of the author.

8 *Eophyton Linnæanum* (Torrell).

9 See Paper on "Footprints and Impressions of Animals," *Am. Journal of Science*, 1873.

10 They probably belong to a large sponge named by Billings *Trichospongia sericea.*

11 *Amphispongia.*

12 *Geological Magazine*, May, 1878.

13 It is regarded as somewhat doubtful whether these are Hydroids or Bryozoa.

14 *Heliopora*, an Alcyonarian; *Pocillopora*, an Anthozoan.

15 *Haplophyllia, Guynia, Duncania*, of Pourtales.

16 *Palæchinus.*

17 Some of the earliest appear to be allies of the modern limpets.

18 "Un produit de l'imagination, sans aucun fondement dans la réalité."

19 *Hymenocaris.*

20 Phyllopods and Ostracods.

21 *Pterygotus, Eurypterus*, &c.

22 Whitfield, *Am. Journ. of Sci.*, 1880.

23 *Report on Devonian Fossil Plants of Canada*, 1871. *Story of the Earth and Man*, 1873. *Address to American Association*, 1875.

24 See the important memoir of Barrande on the Silurian Brachiopods, in which, as the result of the most elaborate and detailed comparisons, he concludes that in the case of these shells, as in that of the Cephalopods and Trilobites, the introduction of species in geological time has not occurred by modification, but must have depended on a creative process. It is such painstaking researches as those of the great Bohemian palæontologist which must finally settle these questions, in so far as geology is concerned.

25 *Geological Magazine*, November, 1869.

26 The genus *Buthotrephis* includes supposed branching sea-weeds of the Silurian. For this reason I would propose the name *Protannularia* for these plants.

27 *Lycopodiaceæ*.

28 Allied to those named by Brongniart *Aetheotesta*.

29 *Cordaites*.

30 Paper by Sir W. Dawson in *Chicago Academy's Bulletin*, 1886.

31 *Calamodendron* and *Arthropitys* are forms of this kind.

32 Grand' Eury and Williamson have directed attention to this in the case of those of France and England.

33 *Amphioxus*.

34 *Petromyzon*, &c.

35 Dr. Newberry has kindly furnished me with specimens, and Dr. Harrington has submitted to analysis portions of shale filled with these little teeth, the result giving 2·58 of calcium phosphate for the whole, which indicates that the Conodonts are really bone. Their microscopic structure approaches to that of the dentine of such Carboniferous fishes as *Diplodus*. Hinde has described Conodonts from the Silurian of Canada.

36 *Ueber Conodonten*: Munich, 1886.

37 *Lepidosteus*.

38 *Palæichthyes* of Günther.

39 *Dinichthys Terrelli* and *D. Hertzeri* (Newberry).

40 Cestracionts.

41 Selachians.

42 *Amphipeltis paradoxus* of Salter.

43 Genus *Strophia*. I have provisionally named the St. John species *Strophites erianus*.

44 The enlarged figure of *Pupa vetusta* is too much elongated, and the aperture is somewhat conjectural, as it is usually crushed.

45 *Dawsonella* of Bradley.

46 *Archiulidæ* of Scudder.

47 *Euphobesia armigera* (Meek and Worthen), from Illinois.

48 About fifty in all, as I learn from Mr. Scudder.

49 *Orthoptera*.

50 *Neuroptera*.

51 *Coleoptera*.

52 *Tincæ*.

53 One highly specialised Carboniferous insect recently found is the *Protophasma* of Brongniart, a relative of the modern "Walking-sticks."

54 This was first described as part of the larva of a Dragon-fly. It is now recognised as belonging to a Scorpion.

55 *Protolycosa* (Roemer).

56 *Menopoma*, *Menobranchus*, &c.

57 *Ophiderpeton Brownriggii*.

58 *Diplichnites*.

59 These are known in some of the smaller species, but not as yet in the larger.

60 *Hylonomus*. See Fig. facing this chapter.

61 *Mastodonsaurus* or *Labyrinthodon*.

62 *Palæosiren Beinertii* of Geinitz.

63 *Hyleopeton*.

64 *Diadictes* and *Bolasaurus* (Cope).

65 *Enaliosauria*, including *Ichthyopterygia* and *Sauropterygia*.

66 *Anomodontia* and *Theriodontia*.

67 *Geology of Oxford*, p. 227.

68 Cope has proposed the names *Camerosaurus*, *Amphicœlias*, &c., for these problematical animals. Marsh names them *Titanosaurus*, *Atlantosaurus*, &c., while Owen holds that some of them at least are identical with his genus *Chondrosteosaurus*. Seeley and Hulke adopt the name *Ornithopsis*, and support Cope's view of their nature.

69 *Ratitæ*.

70 Woodward in a recent paper refers to a still more curious resemblance of the Dinosaurs to the biped lizard of Australia (*Chlamydosaurus*), which runs on its hind limbs, and even perches on trees.

71 A poplar occurs in Greenland, in beds held to be Lower Cretaceous.

72 By some regarded as Upper Cretaceous.

73 First recognized in American Eocene by Newberry.

74 Described by La Harpe and Gaudin, and recently by Gardner.

75 Recent discoveries have since the publication of the first edition removed the Bovey Tracey beds from the Miocene to the Eocene. See Reports of Mr. Starkie Gardner to the British Association.

76 Lyell, *Principles*; Brown, *Florula Discoana*.

77 G. M. Dawson, *Report on 49th Parallel*; *Reports on British Columbia*.

78 Gray's reasoning is based on the extreme view of the Glacial period now prevalent in America, contrary, as it appears to me, to the actual facts; but with limitations it holds good on more moderate views as well.

79 *Geological Magazine*, July, 1887.

80 *Les Enchainements du Monde Animal*.

81 See Frontispiece to this Chapter.

82 For example, *Tillotherium* of the American Eocene, which was as large as a tapir, and in form resembled a bear.

83 Croll, *Climate and Time*.

84 Notes on Post-Pliocene of Canada; *Acadian Geology*, 3rd edition.

85 The actual reason for belief in the past existence of land in the basin of the Indian Ocean is found in the close relationship of forms of life found in Madagascar, Southern Asia, and Australia.

86 Traditions of this animal, a veritable primæval unicorn, are said still to exist in Siberia.

87 As, for instance, those of Cro-Magnon, and Mentone, and Engis.

88 De Puyot and Lohert, Namur, 1887.

89 Religious Tract Society, 1878.

90 May, 1887.

91 *Climate and Time*, a work in which these and other matters relating to the Glacial period are very well discussed.

92 Kimber, quoted by Southall.

93 *Report on Devonian Plants of Canada*, 1871.

94 The true meaning of Hebrews i. 2 and xi. 3.